CW00517978

Classic Papers
in
Control
Theory

EDITED BY
Richard Bellman
& Robert Kalaba

Dover Publications, Inc.
Mineola, New York

Copyright

Copyright © 1964, 1992 by Dover Publications, Inc.
All rights reserved.

Bibliographical Note

This Dover edition, first published in 2017, is a new selection of papers, published for the first time in collected form. The editors and publisher are grateful to the authors and original publishers for permission to reproduce these papers, and to the Directors of the Columbia University Library for assistance in obtaining copies of the articles for reproduction purposes. The present edition was previously published by Dover in 1964 under the title *Selected Papers on Mathematical Trends in Control Theory.*

Library of Congress Cataloging-in-Publication Data

Names: Bellman, Richard, 1920–1984, editor. | Kalaba, Robert E., editor.
Title: Classic papers in control theory / edited by Richard Bellman and
    Robert Kalaba.
Other titles: Selected papers on mathematical trends in control theory
Description: Mineola, New York : Dover Publications, Inc., [2017] | Reprint
    of: Selected papers on mathematical trends in control theory / edited by
    Richard Bellman and Robert Kalaba, [1964]
Identifiers: LCCN 2017022367| ISBN 9780486818566 (paperback) | ISBN
    048681856X (paperback)
Subjects: LCSH: Control theory. | BISAC: TECHNOLOGY & ENGINEERING /
    Electrical.
Classification: LCC QA402.3 .B4 2017 | DDC 629.8/312—dc23
LC record available at https://lccn.loc.gov/2017022367

Manufactured in the United States by LSC Communications
81856X01   2017
www.doverpublications.com

CONTENTS

INTRODUCTION

MAN HAS two principal objectives in the scientific study of his environment: he wants to understand and to control. The two goals reinforce each other, since deeper understanding permits firmer control, and, on the other hand, systematic application of scientific theories inevitably generates new problems which require further investigation, and so on.

It might be assumed that a fine-grained descriptive theory of terrestrial phenomena would be required before an adequate theory of control could be constructed. In actuality, this is not the case, and, indeed, circumstances themselves force us into situations where we must exert regulatory and corrective influences without complete knowledge of basic causes and effects. In connection with the design of experiments, space travel, economics, and the study of cancer, we encounter processes which are not fully understood. Yet design and control decisions are required.

It is easy to see that in the treatment of complex processes, attempts at complete understanding at a basic level may consume so much time and so large a quantity of resources as to impede us in more immediate goals of control.

The mathematical aspects are intriguing. First of all, there are fairly straightforward questions involved in determining optimal control in the presence of complete knowledge of the properties of the underlying physical system. Secondly, there are the more difficult and recondite questions of determining the extent of control that can be exerted, granted only certain pieces of information.

This challenge to the mathematician offered by modern control theory opens new vistas to those who will look, a veritable wilderness of problems for the pioneer, all quite different from the well-plowed classical fields. Nevertheless, the powerful and elegant tools forged in nineteenth-century workshops form the bases for many of the most effective conceptual, analytic, and computational procedures we possess.

The actual history of the mathematical theory is interesting. From Maxwell and Vyshnegradskii to the beginning of World War II, the basic tool was the differential equation, primarily in linear form, with no stochastic overtones. During the war, linear theory and quadratic criteria were still popular since they permitted extensive use of transform techniques and complex variable methods, but due to the efforts of Kolmogorov and Wiener stochastic processes were introduced. From 1945, through the research of the Russian school, Lure, Letov, Pontryagin, and others, and the American school, Booton, Dreyfus, LaSalle, Lefschetz, Minorsky, Kac and Siegert, the editors, and others, nonlinear equations and nonlinear stochastic processes became familiar to the control engineer.

Perhaps the most important mathematical development is this "shotgun wedding" of the classical optimization theory and the classical theory of stochastic processes. The resulting amalgam has already made significant contributions across the scientific board: biology, economics, engineering, psychology.

In the collection of papers offered here we have attempted to follow some natural and logical lines of development from the feedback concept first emphasized by Maxwell and Vyshnegradskii to current work in adaptive control processes. In any such collection it is impossible to present all significant contributions or even to mention a number of fundamental ideas. The papers that have been chosen are in the main connected with aspects of control theory to which we ourselves have been attracted, and are thus able to assess in some degree. The field is one of remarkable proliferation of problems and ideas, as the reader pursuing the references will soon observe.

Starting with feedback control processes involving small deviations from equilibrium, one is led to linear equations and thus to stability problems for linear equations with constant coefficients. A more realistic appraisal leads to nonlinear equations and the theory of Poincaré and Lyapunov. It is then quite natural, yielding to the pressure of practice, to consider, following Minorsky, time lags and thus differential-difference equations. As a prelude to the modern approach which takes account of stochastic phenomena, we turn to linear prediction theory, as envisaged by Kolmogorov and Wiener. Following this, we enter the area of contemporary theory, "bang-bang" control theory in the general format of LaSalle, Gamkrelidze, and others, the Pontryagin maximum principle, and dynamic programming, applied to stochastic and adaptive control.

It will be clear upon reading these papers that control theory is a vital and growing field with enormous promise. Our hope is that this collection of papers will aid young engineers and mathematicians to obtain the overall perspective that is essential for successful research, and thus to help them embark upon their own programs.

It is a great pleasure to acknowledge the many helpful comments and constructive criticism of three old and cherished friends who have themselves contributed so much to control theory, J. P. LaSalle, S. Lefschetz, and N. Minorsky.

RICHARD BELLMAN
ROBERT KALABA

Santa Monica, 1964

ON GOVERNORS*

by J. C. Maxwell

ALTHOUGH isolated examples of control concepts occur throughout recorded history, from the irrigation mentioned in the code of Hammurabi[1] to the centrifugal governor discussed by Huygens for the regulation of wind-mills and water wheels,[2] it was not until the flowering of the English Industrial Revolution that these techniques came into widespread use and became of great significance. The familiar governor of Watt, used for his steam engine,[3] was only one of a number of similar devices which began to play important roles in industry. As a result, serious scientific attention was focused upon the use of governors, regulators, and related mechanisms.

These contrivances are cleverly devised to use the very deviation from desired performance of a system to call upon an actuator to exert a restoring force. This idea must be used with caution, however, since under unfavorable circumstances, the effect which is supposed to be abated can actually be abetted. It appears that Maxwell was the first to realize that these phenomena could be analyzed in mathematical terms, and that there were some subtle mathematical problems contained in the engineering questions. Among the foremost of these is that of the stability of the resulting system.

In the paper that follows, he reviews a number of control devices and transforms the problem of choosing combinations of elements which will yield efficient operation into that of choosing parameters which will produce solutions of a differential equation with desired behavior over time.

Using standard techniques of perturbation theory, Maxwell thus reduces the design problem to that of determining the location of the roots of algebraic polynomials. He resolves it easily enough for quadratic and cubic polynomials, and points out that the problem for nth order polynomials is

---

* From *Proceedings of the Royal Society of London*, Vol. 16, 1868, pp. 270–283.

[1] G. Newton, L. Gould, and J. Kaiser, *Analytical Design of Linear Feedback Control.* New York: John Wiley and Sons, 1957, p. 6.

[2] See the article by H. Bateman, number 2 in our collection, for the reference.

[3] It is interesting to note that the problem of devising an efficient linkage for the conversion of rectilinear into circular motion was responsible for the work of Čebyčev (Tchebychef) on approximation. See the third article in R. Bellman (editor), *A Collection of Modern Mathematical Classics: Analysis.* New York: Dover Publications, 1961.

nontrivial. The general solution was given independently by Routh and Hurwitz; see the paper by Hurwitz.

This paper sets the stage for much of the subsequent development.

Independently, the importance of control theory was recognized by the Russian engineer, Vyshnegradskii. See:

J. Vyshnegradskii, "Sur la théorie générale des régulateurs," *Compt. Rend. Acad. Sci. Paris*, Vol. 83, 1876, pp. 318–321.

———, "Über direkt wirkende Regulatoren," *Der Civilingenieur*, (2), Vol. 23, 1877, pp. 95–132.

Examples of this work are discussed by Pontryagin in his recent book on differential equations.

I. " On Governors." By J. Clerk Maxwell, M.A., F.R.SS.L. & E.
Received Feb. 20, 1868.

A Governor is a part of a machine by means of which the velocity of the machine is kept nearly uniform, notwithstanding variations in the driving-power or the resistance.

Most governors depend on the centrifugal force of a piece connected with a shaft of the machine. When the velocity increases, this force increases, and either increases the pressure of the piece against a surface or moves the piece, and so acts on a break or a valve.

In one class of regulators of machinery, which we may call *moderators* *, the resistance is increased by a quantity depending on the velocity. Thus in some pieces of clockwork the moderator consists of a conical pendulum revolving within a circular case. When the velocity increases, the ball of the pendulum presses against the inside of the case, and the friction checks the increase of velocity.

In Watt's governor for steam-engines the arms open outwards, and so contract the aperture of the steam-valve.

In a water-break invented by Professor J. Thomson, when the velocity is increased, water is centrifugally pumped up, and overflows with a great velocity, and the work is spent in lifting and communicating this velocity to the water.

In all these contrivances an increase of driving-power produces an increase of velocity, though a much smaller increase than would be produced without the moderator.

But if the part acted on by centrifugal force, instead of acting directly on the machine, sets in motion a contrivance which continually increases the resistance as long as the velocity is above its normal value, and reverses its action when the velocity is below that value, the governor will bring the velocity to the same normal value whatever variation (within the working limits of the machine) be made in the driving-power or the resistance.

I propose at present, without entering into any details of mechanism, to direct the attention of engineers and mathematicians to the dynamical theory of such governors.

It will be seen that the motion of a machine with its governor consists in general of a uniform motion, combined with a disturbance which may be expressed as the sum of several component motions. These components may be of four different kinds :—

1. The disturbance may continually increase.
2. It may continually diminish.
3. It may be an oscillation of continually increasing amplitude.
4. It may be an oscillation of continually decreasing amplitude.

The first and third cases are evidently inconsistent with the stability of the motion ; and the second and fourth alone are admissible in a good governor. This condition is mathematically equivalent to the condition that all the possible roots, and all the possible parts of the impossible roots, of a certain equation shall be negative.

I have not been able completely to determine these conditions for equa-

---

* See Mr. C. W. Siemens " On Uniform Rotation," Phil. Trans. 1866, p. 657.

272   Mr. J. C. Maxwell *on Governors*.   [Mar. 5,

tions of a higher degree than the third; but I hope that the subject will obtain the attention of mathematicians.

The actual motions corresponding to these impossible roots are not generally taken notice of by the inventors of such machines, who naturally confine their attention to the way in which it is *designed* to act; and this is generally expressed by the real root of the equation. If, by altering the adjustments of the machine, its governing power is continually increased, there is generally a limit at which the disturbance, instead of subsiding more rapidly, becomes an oscillating and jerking motion, increasing in violence till it reaches the limit of action of the governor. This takes place when the possible part of one of the impossible roots becomes positive. The mathematical investigation of the motion may be rendered practically useful by pointing out the remedy for these disturbances.

This has been actually done in the case of a governor constructed by Mr. Fleeming Jenkin, with adjustments, by which the regulating power of the governor could be altered. By altering these adjustments the regulation could be made more and more rapid, till at last a dancing motion of the governor, accompanied with a jerking motion of the main shaft, showed that an alteration had taken place among the impossible roots of the equation.

I shall consider three kinds of governors, corresponding to the three kinds of moderators already referred to.

In the first kind, the centrifugal piece has a constant distance from the axis of motion, but its pressure on a surface on which it rubs varies when the velocity varies. In the *moderator* this friction is itself the retarding force. In the *governor* this surface is made moveable about the axis, and the friction tends to move it; and this motion is made to act on a break to retard the machine. A constant force acts on the moveable wheel in the opposite direction to that of the friction, which takes off the break when the friction is less than a given quantity.

Mr. Jenkin's governor is on this principle. It has the advantage that the centrifugal piece does not change its position, and that its pressure is always the same function of the velocity. It has the disadvantage that the normal velocity depends in some degree on the coefficient of sliding friction between two surfaces which cannot be kept always in the same condition.

In the second kind of governor, the centrifugal piece is free to move further from the axis, but is restrained by a force the intensity of which varies with the position of the centrifugal piece in such a way that, if the velocity of rotation has the normal value, the centrifugal piece will be in equilibrium in every position. If the velocity is greater or less than the normal velocity, the centrifugal piece will fly out or fall in without any limit except the limits of motion of the piece. But a break is arranged so that it is made more or less powerful according to the distance of the centrifugal piece from the axis, and thus the oscillations of the centrifugal piece are restrained within narrow limits.

Governors have been constructed on this principle by Sir W. Thomson and by M. Foucault. In the first, the force restraining the centrifugal piece is that of a spring acting between a point of the centrifugal piece and a fixed point at a considerable distance, and the break is a friction-break worked by the reaction of the spring on the fixed point.

In M. Foucault's arrangement, the force acting on the centrifugal piece is the weight of the balls acting downward, and an upward force produced by weights acting on a combination of levers and tending to raise the balls. The resultant vertical force on the balls is proportional to their depth below the centre of motion, which ensures a constant normal velocity. The break is :—in the first place, the variable friction between the combination of levers and the ring on the shaft on which the force is made to act; and, in the second place, a centrifugal air-fan through which more or less air is allowed to pass, according to the position of the levers. Both these causes tend to regulate the velocity according to the same law.

The governors designed by the Astronomer Royal on Mr. Siemens's principle for the chronograph and equatorial of Greenwich Observatory depend on nearly similar conditions. The centrifugal piece is here a long conical pendulum, not far removed from the vertical, and it is prevented from deviating much from a fixed angle by the driving-force being rendered nearly constant by means of a differential system. The break of the pendulum consists of a fan which dips into a liquid more or less, according to the angle of the pendulum with the vertical. The break of the principal shaft is worked by the differential apparatus; and the smoothness of motion of the principal shaft is ensured by connecting it with a fly-wheel.

In the third kind of governor a liquid is pumped up and thrown out over the sides of a revolving cup. In the governor on this principle, described by Mr. C. W. Siemens, the cup is connected with its axis by a screw and a spring, in such a way that if the axis gets ahead of the cup the cup is lowered and more liquid is pumped up. If this adjustment can be made perfect, the normal velocity of the cup will remain the same through a considerable range of driving-power.

It appears from the investigations that the oscillations in the motion must be checked by some force resisting the motion of oscillation. This may be done in some cases by connecting the oscillating body with a body hanging in a viscous liquid, so that the oscillations cause the body to rise and fall in the liquid.

To check the variations of motion in a revolving shaft, a vessel filled with viscous liquid may be attached to the shaft. It will have no effect on uniform rotation, but will check periodic alterations of speed.

Similar effects are produced by the viscosity of the lubricating matter in the sliding parts of the machine, and by other unavoidable resistances; so that it is not always necessary to introduce special contrivances to check oscillations.

I shall call all such resistances, if approximately proportional to the velocity, by the name of " viscosity," whatever be their true origin.

In several contrivances a differential system of wheelwork is introduced between the machine and the governor, so that the driving-power acting on the governor is nearly constant.

I have pointed out that, under certain conditions, the sudden disturbances of the machine do not act through the differential system on the governor, or *vice versâ*. When these conditions are fulfilled, the equations of motion are not only simple, but the motion itself is not liable to disturbances depending on the mutual action of the machine and the governor.

Distinction between Moderators and Governors.

In regulators of the first kind, let P be the driving-power and R the resistance, both estimated as if applied to a given axis of the machine. Let V be the normal velocity, estimated for the same axis, and $\frac{dx}{dt}$ the actual velocity, and let M be the moment of inertia of the whole machine reduced to the given axis.

Let the governor be so arranged as to increase the resistance or diminish the driving-power by a quantity $F\left(\frac{dx}{dt}-V\right)$, then the equation of motion will be

$$\frac{d}{dt}\left(M\frac{dx}{dt}\right)=P-R-F\left(\frac{dx}{dt}-V\right). \quad \ldots \ldots (1)$$

When the machine has obtained its final rate the first term vanishes, and

$$\frac{dx}{dt}=V+\frac{P-R}{F}. \quad \ldots \ldots \ldots \ldots \ldots (2)$$

Hence, if P is increased or R diminished, the velocity will be permanently increased. Regulators of this kind, as Mr. Siemens [*] has observed, should be called moderators rather than governors.

In the second kind of regulator, the force $F\left(\frac{dx}{dt}-V\right)$, instead of being applied directly to the machine, is applied to an independent moving piece, B, which continually increases the resistance, or diminishes the driving-power, by a quantity depending on the whole motion of B.

If y represents the whole motion of B, the equation of motion of B is

$$\frac{d}{dt}\left(B\frac{dy}{dt}\right)=F\left(\frac{dx}{dt}-V\right), \quad \ldots \ldots \ldots (3)$$

and that of M

$$\frac{d}{dt}\left(M\frac{dx}{dt}\right)=P-R-F\left(\frac{dx}{dt}-V\right)+Gy, \quad \ldots \ldots (4)$$

where G is the resistance applied by B when B moves through one unit of space.

[*] "On Uniform Rotation," Phil. Trans. 1866, p. 657.

We can integrate the first of these equations at once, and we find

$$B \frac{dy}{dt} = F (x - Vt) ; \quad \cdot \quad \cdot \quad \cdot \quad \cdot \quad \cdot \quad \cdot \quad \cdot \quad \cdot \quad \cdot \quad (5)$$

so that if the governor B has come to rest $x = Vt$, and not only is the velocity of the machine equal to the normal velocity, but the position of the machine is the same as if no disturbance of the driving-power or resistance had taken place.

Jenkin's Governor.—In a governor of this kind, invented by Mr. Fleeming Jenkin, and used in electrical experiments, a centrifugal piece revolves on the principal axis, and is kept always at a constant angle by an appendage which slides on the edge of a loose wheel, B, which works on the same axis. The pressure on the edge of this wheel would be proportional to the square of the velocity ; but a constant portion of this pressure is taken off by a spring which acts on the centrifugal piece. The force acting on B to turn it round is therefore

$$F^1 \overline{\frac{dx}{dt}}\Big|^2 - C^1 ;$$

and if we remember that the velocity varies within very narrow limits, we may write the expression

$$F\left(\frac{dx}{dt} - V_1\right)'$$

where F is a new constant, and V_1 is the lowest limit of velocity within which the governor will act.

Since this force necessarily acts on B in the positive direction, and since it is necessary that the break should be taken off as well as put on, a weight W is applied to B, tending to turn it in the negative direction ; and, for a reason to be afterwards explained, this weight is made to hang in a viscous liquid, so as to bring it to rest quickly.

The equation of motion of B may then be written

$$B\frac{d^2y}{dt^2} = F\left(\frac{dx}{dt} - V_1\right) - Y\frac{dy}{dt} - W, \quad \cdot \quad \cdot \quad \cdot \quad \cdot \quad \cdot \quad (6)$$

where Y is a coefficient depending on the viscosity of the liquid and on other resistances varying with the velocity, and W is the constant weight.

Integrating this equation with respect to t, we find

$$B \frac{dy}{dt} = F (x - V_1 t) - Yy - Wt. \quad \cdot \quad \cdot \quad \cdot \quad \cdot \quad \cdot \quad (7)$$

If B has come to rest, we have

$$x = \left(V_1 + \frac{W}{F}\right) t + \frac{Y}{F}y, \quad \cdot \quad \cdot \quad \cdot \quad \cdot \quad \cdot \quad \cdot \quad (8)$$

or the position of the machine is affected by that of the governor, but the final velocity is constant, and

$$V_1 + \frac{W}{F} = V, \quad \cdot \quad \cdot \quad \cdot \quad \cdot \quad \cdot \quad \cdot \quad \cdot \quad \cdot \quad \cdot \quad (9)$$

where V is the normal velocity.

The equation of motion of the machine itself is

$$M\frac{d^2x}{dt^2}=P-R-F\left(\frac{dx}{dt}-V_1\right)-Gy. \quad \cdots \cdots \quad (10)$$

This must be combined with equation (7) to determine the motion of the whole apparatus. The solution is of the form

$$x=A_1e^{n_1t}+A_2e^{n_2t}+A_3e^{n_3t}+Vt, \quad \cdots \cdots \cdots \quad (11)$$

where n_1, n_2, n_3 are the roots of the cubic equation

$$MBn^3+(MY+FB)n^2+FYn+FG=0. \quad \cdots \cdots \quad (12)$$

If n be a pair of roots of this equation of the form $a\pm\sqrt{-1}b$, then the part of x corresponding to these roots will be of the form

$$e^{at}\cos(bt+\beta).$$

If a is a negative quantity, this will indicate an oscillation the amplitude of which continually decreases. If a is zero, the amplitude will remain constant, and if a is positive, the amplitude will continually increase.

One root of the equation (12) is evidently a real negative quantity. The condition that the real part of the other roots should be negative is

$$\left(\frac{F}{M}+\frac{Y}{B}\right)\frac{Y}{B}-\frac{G}{B}=\text{a positive quantity}.$$

This is the condition of stability of the motion. If it is not fulfilled there will be a dancing motion of the governor, which will increase till it is as great as the limits of motion of the governor. To ensure this stability, the value of Y must be made sufficiently great, as compared with G, by placing the weight W in a viscous liquid if the viscosity of the lubricating materials at the axle is not sufficient.

To determine the value of F, put the break out of gear, and fix the moveable wheel; then, if V and V′ be the velocities when the driving-power is P and P′,

$$F=\frac{P-P'}{V-V'}.$$

To determine G, let the governor act, and let y and y' be the positions of the break when the driving-power is P and P′, then

$$G=\frac{P-P'}{y-y'}.$$

General Theory of Chronometric Centrifugal Pieces.

Sir W. Thomson's and M. Foucault's Governors.—Let A be the moment of inertia of a revolving apparatus, and θ the angle of revolution. The equation of motion is

$$\frac{d}{dt}\left(A\frac{d\theta}{dt}\right)=L, \quad \cdots \cdots \cdots \quad (1)$$

where L is the moment of the applied force round the axis.

Now, let A be a function of another variable ϕ (the divergence of the centrifugal piece), and let the kinetic energy of the whole be

$$\frac{1}{2} A \overline{\frac{d\theta}{dt}}\Big|^2 + \frac{1}{2} B \overline{\frac{d\phi}{dt}}\Big|^2,$$

where B may also be a function of ϕ, if the centrifugal piece is complex.

If we also assume that P, the potential energy of the apparatus, is a function of ϕ, then the force tending to *diminish* ϕ, arising from the action of gravity, springs, &c., will be $\dfrac{dP}{d\phi}$.

The whole energy, kinetic and potential, is

$$E = \frac{1}{2} A \overline{\frac{d\theta}{dt}}\Big|^2 + \frac{1}{2} B \overline{\frac{d\phi}{dt}}\Big|^2 + P. = \int L \, d\theta. \quad \ldots \ldots \quad (2)$$

Differentiating with respect to t, we find

$$\frac{d\phi}{dt}\left(\frac{1}{2}\frac{dA}{d\phi}\overline{\frac{d\theta}{dt}}\Big|^2 + \frac{1}{2}\frac{dB}{d\phi}\overline{\frac{d\phi}{dt}}\Big|^2 + \frac{dP}{d\phi}\right) + A\frac{d\theta}{dt}\frac{d^2\theta}{dt^2} + B\frac{d\phi}{dt}\frac{d^2\phi}{dt^2}$$
$$= L\frac{d\theta}{dt} = \frac{d\theta}{dt}\left(\frac{dA}{d\phi}\frac{d\theta}{dt}\frac{d\phi}{dt} + A\frac{d^2\theta}{dt^2}\right), \quad \bigg\} \quad \cdot \cdot \quad (3)$$

whence we have, by eliminating L,

$$\frac{d}{dt}\left(B\frac{d\phi}{dt}\right) = \frac{1}{2}\frac{dA}{d\phi}\overline{\frac{d\theta}{dt}}\Big|^2 + \frac{1}{2}\frac{dB}{d\phi}\overline{\frac{d\phi}{dt}}\Big|^2 - \frac{dP}{d\phi}. \quad \ldots \ldots \quad (4)$$

The first two terms on the right-hand side indicate a force tending to *increase* ϕ, depending on the squares of the velocities of the main shaft and of the centrifugal piece. The force indicated by these terms may be called the centrifugal force.

If the apparatus is so arranged that

$$P = \tfrac{1}{2} A\omega^2 + \text{const.}, \quad \ldots \ldots \ldots \quad (5)$$

where ω is a constant velocity, the equation becomes

$$\frac{d}{dt}\left(B\frac{d\phi}{dt}\right) = \frac{1}{2}\frac{dA}{d\phi}\left(\overline{\frac{d\theta}{dt}}\Big|^2 - \omega^2\right) + \frac{1}{2}\frac{dB}{d\phi}\overline{\frac{d\phi}{dt}}\Big|^2. \quad \ldots \ldots \quad (6)$$

In this case the value of ϕ cannot remain constant unless the angular velocity is equal to ω.

A shaft with a centrifugal piece arranged on this principle has only one velocity of rotation without disturbance. If there be a small disturbance, the equations for the disturbances θ and ϕ may be written

$$A\frac{d^2\theta}{dt^2} + \frac{dA}{d\phi}\omega\frac{d\phi}{dt} = L, \quad \ldots \ldots \ldots \quad (7)$$

$$B\frac{d^2\phi}{dt^2} - \frac{dA}{d\phi}\omega\frac{d\theta}{dt} = 0. \quad \ldots \ldots \ldots \quad (8)$$

The period of such small disturbances is $\dfrac{dA}{d\phi}(AB)^{-\frac{1}{2}}$ revolutions of the shaft. They will neither increase nor diminish if there are no other terms in the equations.

To convert this apparatus into a governor, let us assume viscosities X and Y in the motions of the main shaft and the centrifugal piece, and a resistance $G\phi$ applied to the main shaft. Putting $\dfrac{dA}{d\phi}\,\omega=K$, the equations

become

$$A\frac{d^2\theta}{dt^2}+X\frac{d\theta}{dt}+K\frac{d\phi}{dt}+G\phi=L, \quad \cdots \cdots \cdots \quad (9)$$

$$B\frac{d^2\phi}{dt^2}+Y\frac{d\phi}{dt}-K\frac{d\theta}{dt} \quad\quad =0. \quad \cdots \cdots \cdots \quad (10)$$

The condition of stability of the motion indicated by these equations is that all the possible roots, or parts of roots, of the cubic equation

$$ABn^3+(AY+BX)n^2+(XY+K^2)n+GK=0 \quad \cdots \cdots \quad (11)$$

shall be negative; and this condition is

$$\left(\frac{X}{A}+\frac{Y}{B}\right)(XY+K^2)>GK. \quad \cdots \cdots \cdots \quad (12)$$

Combination of Governors.—If the break of Thomson's governor is applied to a moveable wheel, as in Jenkin's governor, and if this wheel works a steam-valve, or a more powerful break, we have to consider the motion of three pieces. Without entering into the calculation of the general equations of motion of these pieces, we may confine ourselves to the case of small disturbances, and write the equations

$$\left.\begin{array}{l} A\dfrac{d^2\theta}{dt^2}+X\dfrac{d\theta}{dt}+K\dfrac{d\phi}{dt}+T\phi+J\psi=P-R, \\[2mm] B\dfrac{d^2\phi}{dt^2}+Y\dfrac{d\phi}{dt}-K\dfrac{d\theta}{dt} \quad\quad =0, \\[2mm] C\dfrac{d^2\psi}{dt^2}+Z\dfrac{d\psi}{dt}-T\phi \quad\quad =0, \end{array}\right\} \quad \cdots \cdots \quad (13)$$

where θ, ϕ, ψ are the angles of disturbance of the main shaft, the centrifugal arm, and the moveable wheel respectively, A, B, C their moments of inertia, X, Y, Z the viscosity of their connexions, K is what was formerly denoted by $\dfrac{dA}{d\phi}\omega$, and T and J are the powers of Thomson's and Jenkin's breaks respectively.

The resulting equation in n is of the form

$$\begin{vmatrix} An^2+Xn & Kn+T & J \\ -K & Bn+Y & 0 \\ 0 & -T & Cn^2+Zn \end{vmatrix}=0, \quad \cdots \quad (14)$$

or

$$\left.\begin{array}{l} n^5+n^4\left(\dfrac{X}{A}+\dfrac{Y}{B}+\dfrac{Z}{C}\right)+n^3\left[\dfrac{XYZ}{ABC}\left(\dfrac{A}{X}+\dfrac{B}{Y}+\dfrac{C}{Z}\right)+\dfrac{K^2}{AB}\right] \\[3mm] +n^2\left(\dfrac{XYZ+KTC+K^2Z}{ABC}\right)+n\dfrac{KTZ}{ABC}+\dfrac{KTJ}{ABC}=0. \end{array}\right\} \quad \cdots \quad (15)$$

I have not succeeded in determining completely the conditions of stability of the motion from this equation; but I have found two necessary conditions, which are in fact the conditions of stability of the two governors taken separately. If we write the equation

$$n^5 + pn^4 + qn^3 + rn^2 + sn + t, \quad \ldots \ldots \ldots \quad (16)$$

then, in order that the possible parts of all the roots shall be negative, it is necessary that

$$pq > r \text{ and } ps > t. \quad \ldots \ldots \ldots \quad (17)$$

I am not able to show that these conditions are sufficient. This compound governor has been constructed and used.

On the Motion of a Liquid in a Tube revolving about a Vertical Axis.

Mr. C. W. Siemens's Liquid Governor.—Let ρ be the density of the fluid, k the section of the tube at a point whose distance from the origin measured along the tube is s, r, θ, z the coordinates of this point referred to axes fixed with respect to the tube, Q the volume of liquid which passes through any section in unit of time. Also let the following integrals, taken over the whole tube, be

$$\int \rho k r^2 ds = A, \quad \int \rho r^2 d\theta = B, \quad \int \rho \frac{1}{a} ds = C, \quad \ldots \ldots \quad (1)$$

the lower end of the tube being in the axis of motion.

Let ϕ be the angle of position of the tube about the vertical axis, then the moment of momentum of the liquid in the tube is

$$H = A \frac{d\phi}{dt} + BQ. \quad \ldots \ldots \ldots \quad (2)$$

The moment of momentum of the liquid thrown out of the tube in unit of time is

$$\frac{dH'}{dt} = \rho r^2 Q \frac{d\phi}{dt} + \rho \frac{r}{k} Q^2 \cos a, \quad \ldots \ldots \ldots \quad (3)$$

where r is the radius at the orifice, k its section, and a the angle between the direction of the tube there and the direction of motion.

The energy of motion of the fluid in the tube is

$$W = \frac{1}{2} A \overline{\frac{d\phi}{dt}}^2 + BQ \frac{d\phi}{dt} + \frac{1}{2} CQ^2 . \quad \ldots \ldots \ldots \quad (4)$$

The energy of the fluid which escapes in unit of time is

$$\frac{dW'}{dt} = \rho g Q (h+z) + \frac{1}{2} \rho r^2 Q \overline{\frac{d\phi}{dt}}^2 + \rho \frac{r}{k} \cos a Q^2 \frac{d\phi}{dt} + \frac{1}{2} \frac{\rho}{k^2} Q^3. \quad \ldots \quad (5)$$

The work done by the prime mover in turning the shaft in unit of time is

$$L \frac{d\phi}{dt} = \frac{d\phi}{dt} \left(\frac{dH}{dt} + \frac{dH'}{dt} \right). \quad \ldots \ldots \ldots \quad (6)$$

The work spent on the liquid in unit of time is

$$\frac{dW}{dt} + \frac{dW'}{dt}.$$

Equating this to the work done, we obtain the equations of motion

$$\text{A}\frac{d^2\phi}{dt^2}+\text{B}\frac{d\text{Q}}{dt}+\rho r^2\text{Q}\frac{d\phi}{dt}+\rho\frac{r}{k}\cos a\text{Q}^2=\text{L},\quad\ldots\ldots\quad(7)$$

$$\text{B}\frac{d^2\phi}{dt^2}+\text{C}\frac{d\text{Q}}{dt}+\tfrac{1}{2}\frac{\rho}{k^2}\,\text{Q}^2+\rho g(h+z)-\tfrac{1}{2}\rho\overline{r^2\frac{d\phi}{dt}}^2=0.\quad\ldots\ldots\quad(8)$$

These equations apply to a tube of given section throughout. If the fluid is in open channels, the values of A and C will depend on the depth to which the channels are filled at each point, and that of k will depend on the depth at the overflow.

In the governor described by Mr. C. W. Siemens in the paper already referred to, the discharge is practically limited by the depth of the fluid at the brim of the cup.

The resultant force at the brim is $f=\sqrt{g^2+\omega^4 r^2}$.

If the brim is perfectly horizontal, the overflow will be proportional to $x^{\frac{3}{2}}$ (where x is the depth at the brim), and the mean square of the velocity relative to the brim will be proportional to x, or to $\text{Q}^{\frac{2}{3}}$.

If the breadth of overflow at the surface is proportional to x^n, where x is the height above the lowest point of overflow, then Q will vary as $x^{n+\frac{3}{2}}$, and the mean square of the velocity of overflow relative to the cup as x or as

$$\frac{1}{\text{Q}^{n+\frac{3}{2}}}.$$

If $n=-\tfrac{1}{2}$, then the overflow and the mean square of the velocity are both proportional to x.

From the second equation we find for the mean square of velocity

$$\frac{\text{Q}^2}{k^2}=-\frac{2}{\rho}\Big(\text{B}\frac{d^2\phi}{dt^2}+\text{C}\frac{d\text{Q}}{dt}\Big)+\overline{r^2\frac{d\phi}{dt}}^2-2g(h+r).\quad\ldots\quad(9)$$

If the velocity of rotation and of overflow is constant, this becomes

$$\frac{\text{Q}^2}{k^2}=\overline{r^2\frac{d\phi}{dt}}^2-2g(h+r).\quad\ldots\ldots\ldots\quad(10)$$

From the first equation, supposing, as in Mr. Siemens's construction, that $\cos a=0$ and $\text{B}=0$, we find

$$\text{L}=\rho r^2\text{Q}\frac{d\phi}{dt}.\quad\ldots\ldots\ldots\ldots\quad(11)$$

In Mr. Siemens's governor there is an arrangement by which a fixed relation is established between L and z,

$$\text{L}=-\text{S}z,\quad\ldots\ldots\ldots\ldots\quad(12)$$

whence

$$\frac{\text{Q}^2}{k^2}=\overline{r^2\frac{d\phi}{dt}}^2-2gh+\frac{2g\rho}{\text{S}}r^2\text{Q}\frac{d\phi}{dt}.\quad\ldots\ldots\quad(13)$$

If the conditions of overflow can be so arranged that the mean square of the velocity, represented by $\frac{\text{Q}^2}{k^2}$, is proportional to Q, and if the strength of the spring which determines S is also arranged so that

$$\frac{Q^2}{k^2}=\frac{2g\rho}{S}r^2\omega Q, \qquad \dots \dots \dots \quad (14)$$

the equation will become, if $2gh=\omega^2 r^2$,

$$0=r^2\left(\overline{\frac{d\phi}{dt}}\Big|^2-\omega^2\right)+\frac{2g\rho}{S}r^2 Q\left(\frac{d\phi}{dt}-\omega\right), \quad \dots \dots \quad (15)$$

which shows that the velocity of rotation and of overflow cannot be constant unless the velocity of rotation is ω.

The condition about the overflow is probably difficult to obtain accurately in practice; but very good results have been obtained within a considerable range of driving-power by a proper adjustment of the spring. If the rim is uniform, there will be a *maximum* velocity for a certain driving-power. This seems to be verified by the results given at p. 667 of Mr. Siemens's paper.

If the flow of the fluid were limited by a hole, there would be a *minimum* velocity instead of a maximum.

The differential equation which determines the nature of small disturbances is in general of the fourth order, but may be reduced to the third by a proper choice of the value of the mean overflow.

Theory of Differential Gearing.

In some contrivances the main shaft is connected with the governor by a wheel or system of wheels which are capable of rotation round an axis, which is itself also capable of rotation about the axis of the main shaft. These two axes may be at right angles, as in the ordinary system of differential bevel wheels; or they may be parallel, as in several contrivances adapted to clockwork.

Let ξ and η represent the angular position about each of these axes respectively, θ that of the main shaft, and ϕ that of the governor; then θ and ϕ are linear functions of ξ and η, and the motion of any point of the system can be expressed in terms either of ξ and η or of θ and ϕ.

Let the velocity of a particle whose mass is m resolved in the direction of x be

$$\frac{dx}{dt}=p_1\frac{d\xi}{dt}+q_1\frac{d\eta}{dt}, \qquad \dots \dots \dots \quad (1)$$

with similar expressions for the other coordinate directions, putting suffixes 2 and 3 to denote the values of p and q for these directions. Then Lagrange's equation of motion becomes

$$\Xi\delta\xi+H\delta\eta-\Sigma m\left(\frac{d^2 x}{dt^2}\delta x+\frac{d^2 y}{dt^2}\delta y+\frac{d^2 z}{dt^2}\delta z\right)=0, \quad \dots \quad (2)$$

where Ξ and H are the forces tending to increase ξ and η respectively, no force being supposed to be applied at any other point.

Now putting

$$\delta x=p_1\,\delta\xi+q_1\,\delta\eta, \qquad \dots \dots \dots \quad (3)$$

and

$$\frac{d^2 x}{dt^2}=p_1\frac{d^2\xi}{dt^2}+q_1\frac{d^2\eta}{dt^2}, \qquad \dots \dots \quad (4)$$

the equation becomes

$$\left(\Xi-\Sigma mp^2\frac{d^2\xi}{dt^2}-\Sigma mpq^2\frac{d^2\eta}{dt^2}\right)\delta\xi+\left(\text{H}-\Sigma mpq\frac{d^2\xi}{dt^2}-\Sigma mq^2\frac{d^2\eta}{dt^2}\right)\delta\eta=0; \quad (5)$$

and since $\delta\xi$ and $\delta\eta$ are independent, the coefficient of each must be zero.

If we now put

$$\Sigma(mp^2)=\text{L}, \quad \Sigma(mpq)=\text{M}, \quad \Sigma(mq^2)=\text{N}, \quad \cdots \quad (6)$$

where

$$p^2=p_1^2+p_2^2+p_3^2, \quad pq=p_1q_1+p_2q_2+p_3q_3, \text{ and } q^2=q_1^2+q_2^2+q_3^2,$$

the equations of motion will be

$$\Xi=\text{L}\frac{d^2\xi}{dt^2}+\text{M}\frac{d^2\eta}{dt^2}, \quad \cdots \cdots \cdots \quad (7)$$

$$\text{H}=\text{M}\frac{d^2\xi}{dt^2}+\text{N}\frac{d^2\eta}{dt^2}. \quad \cdots \cdots \cdots \quad (8)$$

If the apparatus is so arranged that $\text{M}=0$, then the two motions will be independent of each other; and the motions indicated by ξ and η will be about conjugate axes—that is, about axes such that the rotation round one of them does not tend to produce a force about the other.

Now let Θ be the driving-power of the shaft on the differential system, and Φ that of the differential system on the governor; then the equation of motion becomes

$$\Theta\delta\theta+\Phi\delta\phi+\left(\Xi-\text{L}\frac{d^2\xi}{dt^2}-\text{M}\frac{d^2\eta}{dt^2}\right)\delta\xi+\left(\text{H}-\text{M}\frac{d^2\xi}{dt^2}-\text{N}\frac{d^2\eta}{dt^2}\right)\delta\eta=0; \quad (9)$$

and if

$$\left.\begin{array}{l}\delta\xi=\text{P}\delta\theta+\text{Q}\delta\phi,\\ \delta\eta=\text{R}\delta\theta+\text{S}\delta\phi,\end{array}\right\} \quad \cdots \cdots \cdots \quad (10)$$

and if we put

$$\left.\begin{array}{l}\text{L}'=\text{LP}^2+2\text{MPR}+\text{NR}^2,\\ \text{M}'=\text{LPQ}+\text{M(PS}+\text{QR})+\text{NRS},\\ \text{N}'=\text{LQ}^2+2\text{MQS}+\text{NS}^2,\end{array}\right\} \quad \cdots \quad (11)$$

the equations of motion in θ and ϕ will be

$$\left.\begin{array}{l}\Theta+\text{P}\Xi+\text{QH}=\text{L}'\dfrac{d^2\theta}{dt^2}+\text{M}'\dfrac{d^2\phi}{dt^2},\\[2mm] \Phi+\text{R}\Xi+\text{SH}=\text{M}'\dfrac{d^2\theta}{dt^2}+\text{N}'\dfrac{d^2\phi}{dt^2}.\end{array}\right\} \quad \cdots \quad (12)$$

If $\text{M}'=0$, then the motions in θ and ϕ will be independent of each other. If M is also 0, then we have the relation

$$\text{LPQ}+\text{NRS}=0; \quad \cdots \cdots \cdots \quad (13)$$

and if this is fulfilled, the disturbances of the motion in θ will have no effect on the motion in ϕ. The teeth of the differential system in gear with the main shaft and the governor respectively will then correspond to the centres of percussion and rotation of a simple body, and this relation will be mutual.

In such differential systems a constant force, H, sufficient to keep the governor in a proper state of efficiency, is applied to the axis η, and the motion of this axis is made to work a valve or a break on the main shaft of the machine. Ξ in this case is merely the friction about the axis of ξ. If the moments of inertia of the different parts of the system are so arranged that $M' = 0$, then the disturbance produced by a blow or a jerk on the machine will act instantaneously on the valve, but will not communicate any impulse to the governor.

THE CONTROL OF AN ELASTIC FLUID*

by H. Bateman

IN the article that follows, Bateman provides a fascinating and invaluable account of control theory based upon linear equations, which is to say upon first order perturbation theory. As later papers will emphasize, these developments of the fundamental engineering concept of feedback control, interesting as they are, miss the full significance of the basic idea of control.

A survey of classical techniques and modern methods is presented in:

R. Bellman, I. Glicksberg, and O. Gross, *Some Aspects of the Mathematical Theory of Control Processes*. The RAND Corporation, R-313, 1958. (Russian translation, Moscow: 1962.)

* ©American Mathematical Society, 1945, all rights reserved, reprinted by permission from the *Bulletin of the American Mathematical Society*, Vol. 51, pp. 601–646.

THE CONTROL OF AN ELASTIC FLUID

H. BATEMAN

1. INTRODUCTION

1.1. Introduction. Mathematicians should pause periodically in their own work and peruse the progress in astronomy, biology, chemistry, economics, engineering, and physics to see if recent advances in these fields suggest problems of mathematical interest. One reason why the Gibbs Lectureship was founded was, indeed, to facilitate a fruitful friendliness between mathematicians and other scientists.

The subject of control is now very important and promises to be so in the future. Much has been written about the control of the air, the control of ships, airplanes, balloons, bombs, gliders, robots and torpedoes. The regulation of rotation became important in the early days of the telescope and steam engine. The related problem of stability is important now for electric motors, marine engines, hydraulic turbines and the generating plants for the distribution of gas and electricity for there is generally an economical speed of operation. In radio telegraphy a certain speed may be needed in order to get a desired frequency.

Controls are necessary in the chemical industries and in mining. They are useful in entertainment and were much needed when arc lights were used for illumination. Fountains which begin to play automatically at sunset are used at exhibitions. Appold's home in London had many automatic devices to interest visitors.

The control of conditions under which observations are made is of great importance to the astronomer, the physicist and the aeronautical engineer. The designer of an engine plans to regulate the flow, pressure, temperature and composition of his working fluid so that the engine will run smoothly and economically.

The control of combustion may be important not only for economical reasons but also to avoid the production of smoke. On the other hand this production may be desirable sometimes when a smoke screen is needed. In such a case there should be flexibility of control. The subject of control is important also in refrigeration, air conditioning and the preparation of food. Great attention is being paid to human comfort. We are in an era of air conditioning on a large scale

The seventeenth Josiah Willard Gibbs lecture delivered at Chicago, Illinois, November 26, 1943, under the auspices of the American Mathematical Society; received by the editors April 2, 1945.

and this requires the solution of many problems of control. It is now understood throughout the land that the provision of the proper atmospheric conditions for the comfort of workmen and the performance of good work is even more important than the regulation of the supply of air and fuel to an engine. Precise weather is needed for precision work and for the manufacture of instruments of precision such as gauges. Proper air conditioning is needed for the production of quality fabrics. The proper temperature must be maintained when stained glass windows are being made. In small arms munition works where dry explosives are handled there is inevitably a certain amount of dust and for safety the amount must be regulated. A gas company must regulate the pressure of gas which it distributes and must also regulate the composition so that an escape of gas may be readily detected by the odour of the escaping gas. Controls are needed for the safety of miners and of workmen in many industries. In the purification of drinking water the rate of supply of chlorine must be regulated.

The subject of control is clearly an enormous one and it is well to bear in mind that advances made in one branch of the subject are sometimes useful in another. A recent aerodynamical torque transmitter for the regulation of a marine engine[1] is based on a principle used in the Remarex carbon dioxide recorder in which there are two pairs of vaned discs, one pair running in air and the other in the flue gas to be tested for CO_2 content. The torque transmitter is intended to prevent the marine engine from racing when, owing to the pitching of the ship, the propeller leaves the water.

Thus advances in marine and aeronautical engineering may depend on advances in chemical engineering. They may depend also on advances in electrical engineering. In wind tunnel research it is important to be able to regulate the velocity of the air moving through the tunnel and one way of doing this has been provided by the extensive work on amplidynes made by the General Electric Company. A large adjustable-speed wind tunnel drive based on the use of amplidynes is described in a paper by Clymer.[2]

In acoustical research it is often necessary to control the vibrations of air in a room. In reverberation work, for instance, the generator of

---

[1] *Aerodynamic marine-engine governor*, Engineering vol. 157 (1944) pp. 447–448.

[2] C. C. Clymer, *Large adjustable-speed wind-tunnel drive*, Transactions of the American Institute of Electrical Engineers vol. 61 (1942) pp. 156–158. Many applications of the amplidyne in closed cycle controllers or regulating systems are described by F. E. Crever, *Fundamental principles of amplidyne applications*, ibid. vol. 62 (1943) pp. 603–606.

sound may be required to produce a pure tone for a certain length of time. If a loud speaker is used and the drive is furnished by an electric current the frequency must be controlled. Sometimes a pure tone is obtained electrically with the aid of an electrical filter. The theory of acoustic and electric filters belongs to the larger subject of the control of vibrations which is important also as it is desirable to eliminate as far as possible the noise and unpleasant vibrations associated with the use of machinery. Much attention has been paid in recent years to the problem of the muffler, the damping of the torsional vibrations of crankshafts and the avoidance of dangerous oscillations in hydraulic transmission lines.

1.2. Quantities which it may be desirable to control.[3]

1. The temperature T.
2. The total density, ρ, or its reciprocal the specific volume.
3. The speed q or the component velocities u, v, w.
4. The mass flow per unit area q.
5. The pressure p.
6. The heat content or enthalpy H.
7. The rate of chemical action, evaporation or condensation.
8. The coefficient of heat transfer.

---

[3] For the control of temperature reference may be made to E. Griffiths, *Thermostats and temperature-regulating instruments*, Griffin, London, 1943; Th. J. Rhodes, *Industrial instruments for measurement and control*, McGraw-Hill, New York and London, 1941; R. L. Weber, *Temperature measurement and control*, Blakiston, Philadelphia, 1941. For the control of various physical quantities see M. Jakob, P. Gmelin and J. Kronert, *Physikalische Kontrolle und Regulierung des Betriebes*, Part I, Leipzig, 1932, Lithoprint, Edwards Brothers, Ann Arbor, Mich., 1943. For matters relating to heat transfer and evaporation see W. H. McAdams, *Heat transmission*, McGraw-Hill, New York, 1942; also A. Fono and C. H. Fielding's papers in Engineering vol. 149 (1940) pp. 79–82. In the theory of hydrogen cooling a mathematical theory of the shaft sealing system is given by D. S. Snell, *The hydrogen-cooled turbine*, Transactions of the American Institute of Electrical Engineers vol. 59 (1940) pp. 35–50. For the recent work on detonation and the physics of flames reference may be made to the book of Bernard Lewis and G. v. Elbe, *Combustion, flames and explosion of gases*, Cambridge, University Press, 1938, and to their paper, *Stability and structure of burner flames*, Journal of Chemistry and Physics vol. 11 (1943) pp. 75–97. For problems of control in the chemical industry reference may be made to the article by H. Seiferheld, *Die Regeltechnik in der chemischen Grossindustrie*, Zeitschrift für Technische Physik, vol. 18 (1937) p. 409 and to a paper by M. Ruhemann, *Equilibrium of liquid and vapour in a rectifying pan*, Physica vol. 4 (1937) pp. 1157–1168, in which use is made of some equations given by F. Bošnjaković. For the electrical precipitation of particles and fumes in gases reference may be made to the work of F. G. Cottrell, *Electrical dust and fume precipitation*, Bulletin of the American Institute of Mining Engineers vol. 67 (1912) pp. 667–675, discussion, pp. 675–680.

9. The velocity of propagation of ignition, detonation or wave motion.

10. The increase or rate of increase of the entropy.

1.3. Types of motion which it may be desirable to produce or avoid.

1. Motions characterized by a certain degree of turbulence such as:

a. Laminar motion in which there is no turbulence.

b. Flow that seems steady but is really turbulent.

c. Tumultuous flow.

d. Swirling flow which is desirable for efficient mixing and good combustion.

e. Flow in which the stream breaks away from the boundary at a selected place.

f. A standard type of turbulence for comparable results in different wind tunnels.

g. Pulsating flow.

h. Flow in which there is a decided rotation about a fixed axis or a mean direction of motion.

2. Motions in which there is a specified relation between p and ρ (barotropic flow) or between ρq and q.

3. Motions accompanied by regular vibrations which may or may not be audible.

a. Regular vibrations may be desirable in experimental work or in some types of engine. Thus in the Kadency engine[4] the intake is timed to occur at the moment when the pressure in the cylinder has fallen below the atmospheric pressure. In the Constantinesco patents regular vibrations in a fluid are used for various types of control in which accurate timing is essential.

b. Both regular and irregular vibrations may be undesirable on account of the noise they produce or because they make a flame unsteady and lead to its extinction or to flash back.

1.4. Methods of control. As the subject of control belongs largely to chemical and electrical engineering only a brief outline of methods can be given here and these will be restricted largely to cases of aerodynamical interest. A few references are given to books from which the reader can obtain information on chemical and electrical methods.

Some of the most important methods of control are:

1. Clever design of fixed boundaries so as to produce desired results with little attention. Thus diffusors, guide vanes and honeycombs may produce the desired type of flow in a wind channel. A

---

[4] S. J. Davies, *An analysis of certain characteristics of a Kadency engine*, Engineering vol. 149 (1940) pp. 515–517, 557–559, 617–620.

spoiler may prevent the flow of air over a roof from injuring the roof. Stationary parts of an engine may be designed to produce a type of flow which is compatible with high efficiency. The wings, body and control surfaces of an airplane may be designed so as to provide low drag and good maneuverability.

2. Devices for altering the form of the boundary of a fluid. Of these the valve is the most important and the design of a suitable valve is often one of the chief steps in the development of a new invention. A list has been formed of nearly eighty different kinds of valves. A valve is generally a device for regulating the rate of flow of a fluid but it may also be used to regulate the pressure or composition of a gas in an enclosure. Mathematically it is usually considered in connection with the regulating device but there are some cases in which equations can be set up for the valve alone. In an attempt to elucidate the action of the throttle valve Joule and Thomson (Lord Kelvin) made their famous porous plug experiment which tests the accuracy of the thermodynamical assumption that the internal energy of a gas depends only on its temperature. Thomson's discussion of the experiment brought into prominence the idea of heat content or enthalpy. Valves may be regarded as including adjustable slots in wings[5] and devices for sucking air from the boundary layer or for blowing air into the boundary layer.[6] The ports of a bunsen burner are also valves. Another device for altering the form of the boundary of the fluid is the fan or blower which produces a forced draught. A modification of this is the windmill or air turbine. A steam turbine or turbine working with gas, mercury vapor or some combination of fluids is another modification.

3. Devices for altering the thermal or electrical condition of a fluid at a boundary. The supply of heat at a boundary or the absorption of heat at a boundary is a most effective way of controlling the motion of a fluid. There are also many electrical devices by means of which an aerodynamic or hydraulic system may be coupled with an electrical system that is furnished with some means of control which may or may not be automatic.

4. Devices for introducing solid particles or liquid in the form of a spray into the body of the working fluid. Gases in the form of jets may also be introduced as in the blowpipe torch and in furnaces. Overfire air jets have been found to be effective for smoke elimination

[5] F. Handley Page, *The Handley-Page wing*, Aeronautical Journal vol. 25 (1921) p. 363.

[6] O. Schrenk, *Boundary layer removal by suction*, National Advisory Committee on Aeronautics, Technical Memorandum No. 974, 1941.

as in a recent paper by Engdahl and Holton.[7] Davis[8] has applied a theory of turbulent air jets developed by Tollmien to the problem of the furnace. The ignition of gaseous mixtures by hot moving particles has been studied by Silver[9] and Paterson.[10] The cooling of gases by sprayed water is one of the methods employed in air conditioning; it has the advantage that the humidity of the air may thereby be controlled at the same time. A history of air conditioning is given by W. H. Carrier.[11] There are cases in which a supply of heat may lead to large fluctuation in temperature. The stability of a simple thermal device which has been called an "academic oven" has been considered by Turner.[12] The analysis depends on a transcendental equation involving both exponential and trigonometrical functions. Oscillations in thermal regulators have been considered also by Himmler.[13]

The problem of stability and of the avoidance of large oscillations becomes important whenever the working fluid is coupled with a mechanical or electrical regulating device. The centrifugal governor invented by Huygens[14] as a possible means of regulating a clock was adapted for windmills and water wheels before it was used by James Watt for the steam engine. A theory of the governor of Huygens has been given by Poor,[15] the theory of Watt's governor and related devices has an extensive literature beginning, perhaps, with the work of

[7] R. B. Engdahl, *Overfire air jets effective for smoke elimination*, Heating, Piping and Air Conditioning, September 1943, p. 481.

[8] R. F. Davis, *The mechanics of flame and air jets*, Engineering vol. 144 (1937) pp. 608–610, 667–668; Proceedings of the Institute of Mechanical Engineers vol. 137 (1938) pp. 11–72.

[9] R. S. Silver, *The ignition of gaseous mixtures by hot particles*, Philosophical Magazine (7) vol. 23 (1936) pp. 633–657.

[10] S. Paterson, *The ignition of inflammable gases by hot moving particles*, Philosophical Magazine (7) vol. 28 (1939) pp. 1–23.

[11] W. H. Carrier, *Air conditioning*, Encyclopedia Brittanica; see also W. H. Carrier, R. E. Cherne and W. A. Grant, *Modern air conditioning, heating and ventilation*, Chicago, Pitman, 1940; C. O. Mackey, *Air conditioning principles*, Scranton, Pa., 1941. Lord Kelvin is credited with a proposal for the use of mechanical cooling as a means of improving human comfort. William Appold devised apparatus for control of temperature and humidity. See J. P. Gassiot, *On Appold's apparatus for regulating temperature and keeping the air in a building at any desired degree of moisture*, Proc. Roy. Soc. London vol. 15 (1867) pp. 144–146.

[12] L. B. Turner, *Self-oscillation in a retroacting thermal conductor*, Proc. Cambridge Philos. Soc. vol. 32 (1936) pp. 663–675.

[13] C. Himmler, *Die Pendelungen bei warmetechnischen Regelvorgängen*, Zeitschrift für Technische Physik vol. 11 (1929) pp. 579–584.

[14] C. Huygens, *Horologii oscillatorii*, Part 5, Paris, 1673; *Horologium*, 1658.

[15] V. C. Poor, *The Huygens governor*, Amer. Math. Monthly vol. 32 (1925) pp. 115–121.

Navier and Poncelet.[16] For a survey of this literature reference may be made to the article of von Mises and to the book of Tolle.[17]

2. THE ALGEBRAIC PROBLEM

2.1. Conditions of stability. In direct regulation the stability of the dynamical, electrical or hydraulic regulating device often can be discussed by the method of small oscillations. The system is generally a compound one which is partly of one type and partly of another. For instance, in the case of the steam engine, the compound system consists of the steam, the valve and the centrifugal governor and so the processes taking place are described by a system of differential equations. In indirect regulation the system is again compound. When a deviation from the norm (or rated value of the quantity to be controlled) passes out of the region of insensitivity, the indicator actuates a motor through the amplifier and a disturbance is produced which tends to annul the disturbance shown by the indicator. When the transient force is no longer operative, the most desirable type of motion of the system is a damped oscillation or a simple decay without oscillation such as is sometimes produced when a jet is used to control the speed of rotation as in Michelson's[18] measurements of the velocity of light by means of a revolving mirror. In the so-called exact regulation the indicator returns to the normal setting after a single swing past it. It is generally, but not always, advantageous to eliminate all the variables but one which may be denoted by x, then, if $D \equiv d/dt$

$$(p_0 D^n + p_1 D^{n-1} + \cdots + p_n)x = f(t).$$

The transient function $f(t)$ may be different from zero only for $0 < t < T$. Then, for $t > T$, x is a sum of terms satisfying the equation with $f(t)$ replaced by zero but it is not certain that all possible solutions of the homogeneous equation enter into the expression for the particular quantity x.

---

[16] J. V. Poncelet, *Cours de méchanique, appliquée aux machines*, Cours de l'école d'application de Metz, 1826. R. v. Mises, *Dynamische Probleme der Maschinenlehre*, Encyklopädie Mathematischen Wissenshaften vol. 4, part 10, pp. 153–355.

[17] M. Tolle, *Die Regelung von Kraftmaschinen*, 3d ed., Berlin, 1921. Tolle gives in particular a theory for the combination of a centrifugal governor and a relay. The theory is presented and amplified for the case of two relays and the effect of the steam by G. W. Higgs-Walker, *Some problems connected with steam turbine governing*, Proceedings of the Institute of Mechanical Engineers vol. 146 (1941) pp. 117–125.

[18] A. A. Michelson, *Measurements of the velocity of light between Mount Wilson and Mount San Antonio*, Astrophysical Journal vol. 65 (1927) pp. 1–14; A. A. Michelson, F. G. Pease and F. Pearson, ibid. vol. 82 (1935) pp. 26–61.

A regulator is said to be stable when, after a transient disturbance, the indicator returns to the region of insensitivity and does not go beyond this. In the case when the free motion involves an undamped or growing motion, the regulator is said to hunt or to be unstable. A sufficient condition for stability is that the roots of the algebraic equation

$$(2.1) \qquad F(z) \equiv p_0 z^n + p_1 z^{n-1} + \cdots + p_n = 0$$

should have only negative real parts. This condition may not be quite necessary because it may happen that a root with positive or zero real part does not happen to give a term in the expression for x. This possibility must be considered because something of an analogous nature seems to occur in some cases when the existence of a double root might make the sufficiency of the foregoing criterion seem doubtful.

The mathematical theory of stability based on the theory of small oscillations may be hard to use on account of lack of knowledge of the constants of the dynamical or electrical system. These can be estimated in many cases as in the theory of airplane stability but it is wise to have means of checking the results or of obtaining results when the computations are too difficult.

An oscillograph for the analysis of governor performance was built by J. E. Allen,[19] and the East Pittsburgh Research Laboratories have built an instrument for analyzing governor performance which satisfies the specifications that have been laid down. Instruments of this nature have been made elsewhere.[20] Dougill has devised an instrument for testing regulators in operation and has used it to test the governors in the gas works.[21]

2.2. **Pseudo-negative roots.** Liénard[22] calls a quantity pseudo-negative when its real part is negative. The criterion for pseudo-negative roots of an algebraic equation is a special case of the criterion that the roots of an algebraic equation should lie within a specified circle in the complex plane, a line being regarded as a degenerate circle. A line may also be transformed into a circle by means of a transformation

---

[19] J. E. Allen, *Oscillograph analyses governor performance*, Power vol. 78 (1934) pp. 610–612.

[20] W. O. Oebon, *A turbine governor performance analyzer*, American Institute of Electrical Engineers vol. 69 (1941) pp. 963–967.

[21] G. Dougill, *Retort house and exhauster governing of gas works*, Engineering vol. 144 (1937) p. 697.

[22] A. Liénard, *Signe de la partie reélle des racines d'une équation algébrique*, J. Math. Pures Appl. (9) vol. 15 (1936) pp. 235–250.

$$Z = (Az + B)/(Cz + D)$$

which does not change the degree of the equation

Cauchy[23] devised a method for finding the criterion and Hermite[24] carried the analysis much further considering particularly the conditions that the roots should lie in the upper half of the complex plane. He used a symmetrical polynomial

$$H(z', z) = i[F(z')F_0(z) - F(z)F_0(z')]/(z - z')$$

associated with the function $F(z)$ and an associated function $F_0(z)$ derived from $F(z)$ by changing i into $-i$ in all the coefficients. The decomposition into squares of an associated quadratic form then indicated the number of pseudo-negative roots, this number being dependent in fact on the signature of the quadratic form.

At a meeting of the London Mathematical Society[25] in 1868, James Clerk Maxwell asked if any member present could point out a way of determining in what cases all the possible parts of the imaginary roots of an algebraic equation are negative. He said that in studying the motion of certain governors for regulating machinery he had found that the stability of the motion depended on this condition, which is easily obtained for a cubic, but becomes difficult in the higher degrees. W. K. Clifford said in reply that by forming an equation whose roots are the sums of the roots of the original equation taken in pairs and by determining the condition of the real roots of this equation being negative, we should obtain the condition required.

Routh[26] used Clifford's idea when formulating conditions for a quartic equation with real coefficients,

$$(2.2) \qquad F(z) = az^4 + bz^3 + cz^2 + dz + e = 0.$$

He says: "Let us form that symmetrical function of the roots which is the product of the sums of the roots taken two and two. If this be called X/a^3, we find $X = bcd - ad^2 - eb^2$. Suppose we know the roots to be imaginary, say $\alpha \pm ip$, $\beta \pm iq$. Then

$$X/a^3 = 4\alpha\beta[(\alpha + \beta)^2 + (p + q)^2][(\alpha + \beta)^2 + (p - q)^2].$$

[23] A. Cauchy, *Calcul des indices des fonctions*, J. École Polytech. vol. 15 (1837) pp. 176–229, *Oeuvres* (2), vol. 1, pp. 416–466.

[24] C. Hermite, *Extrait d'une lettre, Sur le nombre des racines d'une équation algébrique compris des limites données*, J. Reine Angew. Math. vol. 52 (1856) pp. 39–51.

[25] See the discussion of the paper by J. J. Walker, *On the anharmonic sextic*, Proc. London Math. Soc. (1) vol. 2 (1868) pp. 60–61.

[26] E. J. Routh, *Rigid dynamics*, vol. 2, 1897, pp. 192–193; *A treatise on the stability of motion*, London, 1877; *Advanced rigid dynamics*, 6th ed., 1907, pp. 256–307.

Thus, $\alpha\beta$ always takes the sign of X/a and $\alpha+\beta$ always takes the sign of $-b/a$. The signs of both α and β can therefore be determined; and if a, b, X have the same sign, the real parts of the roots are all nega-tive." Routh also formed the equation $G(z)=0$ whose roots are the sums in pairs of the roots of $F(z)=0$ and in the case when the coeffi-cients p_r in the general equation (2.1) are all real he came to the fol-lowing conclusion:

In order that (2.1) may have all its roots pseudo-negative, it is necessary and sufficient that the equations $F(z)=0$, $G(z)=0$ should be complete with coefficients all of one sign. This means that no p should be zero and that if $p_0>0$ then $p_r>0$. If, moreover, the co-efficients of $G(z)$ are q_r, $r=0$, 1, $\cdots$, $n^2/2-n/2$ then if $q_0>0$, we should also have $q_r>0$. These conditions give $n^2/2+n/2$ inequalities while the expected number of conditions is only n so the foregoing conditions are not all independent.

In his work on governors in which he considered particularly the governors designed by Foucault and Lord Kelvin, Maxwell found that the stability depended upon the conditions for pseudo-negative roots of an equation of the fifth degree $z^5+pz^4+qz^3+rz^2+sz+t=0$. He found the necessary conditions $pq>r$, $ps<t$ but could not prove that these conditions were sufficient.

The simplicity of these conditions suggested that there might be necessary and sufficient conditions in the general case which could be formulated in a simple form. The subject of the stability of motion was soon afterwards proposed as a subject for the Adams Prize at the University of Cambridge and the prize was won by E. J. Routh of Peterhouse. He made use of the methods of Cauchy and Charles Sturm and a set of test functions was formed by a cascade process. Writing

$$F(z) = E(z) + O(z) = A(z^2) + zB(z^2)$$

so as to resolve $F(z)$ into its even and odd parts, he made use of the functions $s_0(y)$, $s_1(y)$, $s_2(y)$, $\cdots$, $s_n(y)$, where

$$s_0(y) = p_0y^n - p_2y^{n-1} + \cdots , \quad s_1(y) = p_1y^{n-1} - p_3y^{n-3} + \cdots ,$$

$s_2(y)$ is the remainder with sign changed when use is made of the process for finding the G.C.M. of $s_0(y)$ and $s_1(y)$, $s_3(y)$ is derived from $s_1(y)$ and $s_2(y)$ in a similar way, and so on. It is then clear as in Sturm's theorem that when $s_r(y)=0$, $s_{r+1}(y)$ and $s_{r-1}(y)$ have opposite signs. If E denotes the excess of the number of changes of sign from $+$ to $-$ in $s_0(y)/s_1(y)$ over that from $-$ to $+$, then by Cauchy's theorem the whole number of radical points on the positive side of the axis of y is

$(n+E)/2$. If $E = -n$ the roots are all pseudo-negative. The roots of the equations $A(x) = 0$, $B(x) = 0$ are in this case all negative and occur alternately. In the case of the biquadratic

$$F(z) = z^4 + pz^3 + qz^2 + rz + s = (z^2 + x_1)(z^2 + x_2) + pz(z^2 + x_0)$$

$q = x_1 + x_2$, $r = px_0$, $s = x_1 x_2$ and Routh's test functions are

$$s, \quad p, \quad pq - r = p(x_1 + x_2 - x_0), \quad r(pq - r) - p^2 s = p^2(x_2 - x_0)(x_0 - x_1).$$

The third test function is positive when and only when x_0 lies between x_1 and x_2. The test functions p and $pq - r$ are both positive when $p > 0$ and $x_1 + x_2 - x_0$ is positive. Now if $x_1 - x_0$ is negative, x_2 must be positive. When s is positive, x_1 and x_2 must be either both positive or both negative, hence if x_2 is positive so also is x_1. Routh's criteria for pseudo-negative roots imply then that x_0, x_1, x_2 are all positive and that x_0 lies between x_1 and x_2. When these conditions are all satisfied s is positive, $r(pq - r) - p^2 s$ is positive, $r(pq - r)$ and $q - r/p$ are positive. The ratio r/p of the two last quantities is positive and so q must be positive. If p is also positive r is positive. Routh's conditions are all satisfied and the equation $F(z) = 0$ has pseudo-negative roots. Liénard attributes this converse theorem to E. Jouguet and says that Chipart has extended it to an equation of any degree. A general proof is given in Liénard's paper.

Related polynomials such as $A(-x)$, $B(-x)$ occur naturally in Rayleigh's theory of the driving point reaction in dynamics and in the theory of electric filters.

About 1893 the Swiss engineer Aurel Stodola[27] investigated the stability of regulating devices for turbines, particularly those used in hydroelectric plants. He referred to Thomson and Tait's *Natural philosophy*[28] for the relation between stability and pseudo-negative roots and on this account Corral[29] has called the question of pseudo-negative roots the problem of Lord Kelvin. Previously[30] he had followed Orlando[31] in calling it the problem of Hurwitz because

---

[27] A. Stodola, *Über die Regulierung von Turbinen*, Schweizerische Bauzeitung vol. 22 (1893) pp. 113–117, 121–122, 126–128, 134–135; vol. 23 (1894) pp. 108–112, 115–117.

[28] W. Thomson (Lord Kelvin) and P. G. Tait, *Natural philosophy*, vol. 1, 1879, p. 39.

[29] J. J. Corral, *Nueva solucion del problema de Lord Kelvin sobre ecuaciones de coeficientes reales*, Revista de la Real Academia de Ciencias Exactas, Fisicas y Naturales de Madrid vol. 22 (1928) pp. 25–31.

[30] J. J. Corral, *Nuevos teoremas que resuelven el problema de Hurwitz*, Madrid, Imprenta Clasica Española, 1921.

[31] L. Orlando, *Sul problema di Hurwitz*, Rendiconti Accademia Lincei (5) vol. 19 (1910) pp. 801–805; Math. Ann. vol. 71 (1911) pp. 233–245.

Stodola's compatriot Adolf Hurwitz[32] had investigated the subject and had succeeded in obtaining criteria in the determinantal form

$$p_0 > 0, \quad p_1 > 0, \quad \begin{vmatrix} p_1 & p_0 \\ p_3 & p_2 \end{vmatrix} > 0, \quad \begin{vmatrix} p_1 & p_0 & 0 \\ p_3 & p_2 & p_1 \\ p_5 & p_4 & p_3 \end{vmatrix} > 0,$$

$$\begin{vmatrix} p_1 & p_0 & 0 & 0 \\ p_3 & p_2 & p_1 & p_0 \\ p_5 & p_4 & p_3 & p_2 \\ p_7 & p_6 & p_5 & p_4 \end{vmatrix} > 0, \quad \text{and so on.}$$

The equivalence of the conditions of Routh and Hurwitz was shown by Bompiani,[33] and Orlando obtained a proof by induction of the necessity and sufficiency of Hurwitz's conditions.

The study of equations with complex coefficients is also useful as there are some cases in which the conditions of Routh and Hurwitz are not the simplest possible conditions for pseudo-negative roots. In Appell's *Mécanique rationnelle* a discussion is given of dynamical equations such as

$$x'' + p_1 x' + q_1 x - p_2 y' - q_2 y = 0,$$
$$y'' + p_1 y' + q_1 y + p_2 x' - q_2 x = 0,$$

in which there are gyrostatic terms. These equations are essentially those considered by Sir Horace Lamb in his work on kinetic stability[34] and by E. Jouguet in his work on secular stability.[35] The algebraic equation obtained in the usual way is

$$(z^2 + p_1 z + q_1)^2 + (p_2 z + q_2)^2 = 0$$

but if we put $x + iy = Z$, as Lamb does, there is a single dynamical equation

$$Z'' + (p_1 + ip_2)Z' + q_1 + iq_2 = 0$$

[32] A. Hurwitz, *Über die Bedingungen, unter welchen eine Gleichung nur Wurzeln mit negativen reellen Theilen besitzt,* Math. Ann. vol. 46 (1895) pp. 273–284; *Werke,* vol. 2, pp. 533–545.

[33] E. Bompiani, *Sulle condizioni sotto le quali un equazione a coefficienti reale ammette solo radici con parte reale negative,* Giornale di Matematica vol. 49 (1911) pp. 33–39.

[34] H. Lamb, *On kinetic stability,* Proc. Roy. Soc. London. Ser. A. vol. 80 (1908) pp. 168–177.

[35] E. Jouguet, *Sur la stabilité séculaire quand les forces positionnelles n'admettent pas de potentiel,* C. R. Acad. Sci. Paris vol. 207 (1938) pp. 267–270.

which gives rise to an algebraic equation

$$z^2 + (p_1 + ip_2)z + q_1 + iq_2 = (z + x_1 + iy_1)(z + x_2 + iy_2) = 0.$$

The conditions for pseudo-negative roots are now

$$x_1 + x_2 > 0 \quad \text{and} \quad x_1 x_2 [(x_1 + x_2)^2 + (y_1 - y_2)^2] > 0$$

while the corresponding conditions derived by considering the bi-quadratic equation are $x_1 + x_2 > 0$ and

$$x_1 x_2 [(x_1 + x_2)^2 + (y_1 - y_2)^2][(x_1 + x_2)^2 + (y_1 + y_2)^2] > 0$$

and there is an extra factor in the expression used for the second criterion.

The conditions for the quadratic may be expressed in terms of the quantities I_1, I_2, I_3 which are invariant when the equation is changed into a new equation by a substitution of the form $z = Z + ia$, where a is real. If

$$(Z + ia)^2 + (p_1 + ip_2)(Z + ia) + q_1 + iq_2$$
$$\equiv Z^2 + (P_1 + iP_2)Z + Q_1 + iQ_2$$

then $P_1 = p_1$, $P_2 = p_2 - 2a$, $Q_1 = q_1 + ap_2 - a^2$, $Q_2 = q_2 - ap_1$ and so there are 3 invariants

$$I_1 = P_1 = p_1, \qquad I_2 = Q_1 + P_2^2/4 = q_1 + p_2^2/4,$$
$$I_3 = Q_2 - P_1 P_2/2 = q_2 - p_1 p_2/2.$$

If, in particular, we choose a so that $P_2 = 0$, the equation takes the simple form

$$Z^2 + I_1 Z + I_2 + iI_3 = 0.$$

If Z_1, Z_2 are the roots of this equation and if T_1, T_2 are the roots of the conjugate equation

$$T^2 + I_1 T + I_2 - iI_3 = 0,$$

the equation whose roots are $Z_1 + T_1$, $Z_2 + T_2$, $Z_1 + T_2$, $Z_2 + T_1$ is

$$S^4 + 4I_1 S^3 + (5I_1^2 + 4I_2)S^2 + (2I_1^3 + 8I_1 I_2)S + 4I_1^2 I_2 - 4I_3^2 = 0.$$

This is also the equation whose roots are $z_1 + t_1$, $z_2 + t_2$, $z_1 + t_2$, $z_2 + t_1$ where z_1, z_2 are the roots of the original equation and t_1, t_2 are the roots of its conjugate equation. It should be noticed that the terms in the equation for S involve the invariants and S only, moreover, by using two of these terms expressions

$$x_1 + x_2 = I_1, \qquad x_1 x_2 [(x_1 + x_2)^2 + (y_1 - y_2)^2] = 4(I_1^2 I_2 - I_3^2)$$

are obtained for the quantities that furnish criteria for the roots to be pseudo-negative.

The quadratic equation may be reduced to a canonical form

$$\frac{1}{z + iw_1} + \frac{1}{z + iw_2} + \frac{1}{z + u + iv} = 0$$

where w_1, w_2, u and v are real quantities. The equation is then of stable type (with pseudo-negative roots) when $u > 0$ for the roots are those of the derived function of the cubic

$$(z + iw)(z + iw)(z + u + iv) = 0$$

and so by the theorem of Gauss[36] and Lucas[37] lie within the triangle formed by the points $z = -iw_1$, $z = -iw_2$, $z = -u - iv$ in the complex z-plane. The roots are in fact the foci of the ellipse which touches the sides of this triangle at its middle points.

The equations for determining u, v, w_1, w_2 are

$$3p_1 = 2u, \qquad 3p_2 = 2(v + w_1 + w_2), \qquad 3q_1 = - w_1 w_2 - v(w_1 + w_2),$$

$$3q_2 = u(w_1 + w_2)$$

and so $w_1 + w_2 = 2q_2/p_1$, $w_1 w_2 = -3q_1 - (3p_1 p_2 q_2 - 4q_2^2)/p_1^2$. The quantities w_1, w_2 are thus the roots of the quadratic equation

$$p_1^2 w^2 - 2q_2 p_1 w + 4q_2^2 - 3p_1 p_2 q_2 - 3p_1^2 q_1 = 0$$

which has real roots when $p_1^2 q_1 + p_1 p_2 q_2 - q_2^2 > 0$ or $I_1^2 I_2 - I_3^2 > 0$. When this condition is satisfied the sign of u is positive when $I_1 > 0$.

The extension of Clifford's method which was used for the quadratic may be applied also to the cubic

$$z^3 + (p_1 + ip_2)z^2 + (q_1 + iq_2)z + r_1 + ir_2 = 0.$$

The 5 invariants are $I_1 = p_1$, $3I_2 = p_2^2 + 3q_1$, $3I_3 = 3q_2 - 2p_1 p_2$, $3I_4 = 9r_1 - 3p_1 q_1 + 3p_2 q_2 - 2p_1^2 p_2^2$, $27I_5 = 27r_2 - 9p_2 q_1 - 2p_2^3$. When a substitution $z = Z + ia$ is chosen so that in the new equation the coefficient of Z^2 is real, the new equation is

$$Z^3 + I_1 Z^2 + (I_2 + iI_3)Z + (I_4 + I_1 I_2)/3 + iI_5 = 0.$$

---

[36] C. F. Gauss, *Oeuvres*, vol. 3, 1886, p. 112; vol. 8, 1900, p. 32.

[37] F. Lucas, *Géometrie des polynômes*, J. École Polytech. vol. 29 (1879) pp. 1–33. See also M. Marden, *The location of the zeros of the derivative of a polynomial*, Amer. Math. Monthly vol. 42 (1935) pp. 277–286.

If the roots of this equation are Z_1, Z_2, Z_3 and those of the conjugate equation T_1, T_2, T_3 the equation for $S = Z + T$ may be readily found by elimination and its roots are the 9 quantities $Z_1 + T_1$, $Z_1 + T_2$, $Z_1 + T_3$, $Z_2 + T_1$, $Z_2 + T_2$, $Z_2 + T_3$, $Z_3 + T_1$, $Z_3 + T_2$, $Z_3 + T_3$. The product of the roots of this equation is particularly interesting as it furnishes a quantity

$$K = J(I_1 I_2 - J)^2 - 3J I_1 I_3 I_5 + I_1^2 I_2 I_3 I_5 - J I_2 I_3^2 - I_1^3 I_5^2,$$
$$J = I_4 + I_1 I_2,$$

which is positive when all the roots are pseudo-negative. The necessary and sufficient conditions for pseudo-negative roots are $I_1 > 0$, $K > 0$, and $I > 0$ where these quantities are such that when positive they imply that $X_1 + X_2 + X_3$, $X_1 X_2 X_3$ and $X_2 X_3 + X_3 X_1 + X_1 X_2$ are all positive. To find the invariant I it is helpful to use the notation

$$U_1 = X_1 + iY_1, \quad U_2 = X_2 + iY_2, \quad U_3 = X_3 + iY_3, \quad V_1 = X_1 - iY_1,$$
$$V_2 = X_2 - iY_2, \quad V_3 = X_3 - iY_3, \quad H_{rs} = U_r + V_s.$$
$$\begin{aligned} I &= (H_{22}H_{33} + H_{33}H_{11} + H_{11}H_{22})(H_{23}H_{31} + H_{31}H_{12} \\ &\quad + H_{12}H_{23})(H_{32}H_{13} + H_{13}H_{21} + H_{21}H_{32}) \\ &= 4(X_2 X_3 + X_3 X_1 + X_1 X_2) \left| H_{23}H_{31} + H_{31}H_{12} + H_{12}H_{23} \right|^2. \end{aligned}$$

The quantities U_1, U_2, U_3 are identical with $-Z_1$, $-Z_2$ and $-Z_3$; also

$$I = (W - W_1)(W - W_2)(W - W_3)$$

where

$$\begin{aligned} W_1 &= U_2 U_3 + U_3 U_1 + U_1 U_2 + V_2 V_3 + V_3 V_1 + V_1 V_2 \\ &\quad + (U_1 + U_2 + U_3)(V_1 + V_2 + V_3), \\ W_1 &= U_1 V_1 + U_2 V_2 + U_3 V_3, \qquad W_2 = U_2 V_3 + U_3 V_1 + U_1 V_2, \\ W_3 &= U_3 V_2 + U_1 V_3 + U_2 V_1. \end{aligned}$$

These quantities W_1, W_2, W_3 and a second set of quantities W_1', W_2', W_3' obtained by changing the cyclic order of V_1, V_2, V_3 to V_3, V_2, V_1 are the roots of a sextic equation

$$(W^3 - AW^2 + BW - C)^2 = d^2 DD'$$

where D is the discriminant of the equation for U and D' is the discriminant of the equation for V. This form is indicated by the fact that when the equation for U_1, U_2, U_3 has equal roots or when the equation for V_1, V_2, V_3 has equal roots the two sets of three sums become the same. Also we have identically $W_1' + W_2' + W_3' = W_1 + W_2$

$+W_3, \ W'_2 \, W'_3 + W'_3 \, W'_1 + W'_1 \, W'_2 = W_2 W_3 + W_3 W_1 + W_1 W_2.$

To find d we put $W=0$ and note that $W_1 W_2 W_3 - W'_1 \, W'_2 \, W'_3 = 2d(DD')^{1/2}$. But

$$W_1 W_2 W_3 - W'_1 \, W'_2 \, W'_3$$
$$= (U_2 - U_3)(U_3 - U_1)(U_1 - U_2)(V_2 - V_3)(V_3 - V_1)(V_1 - V_2) = (DD')^{1/2},$$

consequently $2d = 1$. We also have the relation

$$2C = W_1 W_2 W_3 + W'_1 \, W'_2 \, W'_3$$
$$= U_1 U_2 U_3 (V_1^3 + V_2^3 + V_3^3) + V_1 V_2 V_3 (U_1^3 + U_2^3 + U_3^3) + 6 U_1 U_2 U_3 V_1 V_2 V_3$$
$$+ (U_2^2 U_3 + U_3^2 U_2 + U_3^2 U_1 + U_1^2 U_3 + U_1^2 U_2 + U_2^2 U_1)(V_2^2 V_3 + V_3^2 V_2$$
$$+ V_3^2 V_1 + V_1^2 V_3 + V_1^2 V_2 + V_2^2 V_1)$$
$$= R(P'^3 - 3 P' Q' + 3 R') + R'(P^3 - 3 PQ + 3R) + 6 R R'$$
$$+ (PQ - 3R)(P' Q' - 3 R')$$

where P, Q, R are the coefficients in the equation for Z_1, Z_2, Z_3, P', Q', R' the coefficients in the equation for T_1, T_2, T_3.

Hence $2C = R P'^3 + R' P^3 - 6 P' Q' R - 6 PQR' + PQP'Q' + 18 R R'$. Also

$$B = W_2 W_3 + W_3 W_1 + W_1 W_2 = (U_1^2 + U_2^2 + U_3^2)(V_2 V_3 + V_3 V_1 + V_1 V_2)$$
$$+ (U_2 U_3 + U_3 U_1 + U_1 U_2)(V_2 V_3 + V_3 V_1 + V_1 V_2 + V_1^2 + V_2^2 + V_3^2)$$
$$= Q'(P^2 - 2Q) + Q(P'^2 - Q') = P^2 Q' + P'^2 Q - 3 Q Q'.$$

Hence

$$(W - W_1)(W - W_2)(W - W_3) = W^3 - PP'W^2 + (P^2 Q' + P'^2 Q - 3 Q Q')W$$
$$- (R P'^3 + R' P^3 - 6 P' Q' R - 6 PQR' + PQP'Q' + 18 R R')/2 - (DD')^{1/2}/2.$$

With the value $W = Q + Q' + PP'$ the expression for I is

$$I = (Q + Q' + PP')^3 - PP'(Q + Q' + PP')^2$$
$$+ (P^2 Q' + P'^2 Q - 3 Q Q')(Q + Q' + PP')$$
$$- (R P'^3 + R' P^3 - 6 P' Q' R - 6 PQR' + PQP'Q' + 18 R R')/2 - (DD')^{1/2}/2.$$

This is the expression obtained in a former paper.[38] It is well known that

$$D = P^2 Q^2 - 4 P^3 R + 18 PQR - 27 R^2$$

while D' can be expressed in a similar way in terms of P', Q', R'. It should be mentioned that the conditions for pseudo-negative roots for

---

[38] H. Bateman, *Stability of the parachute and helicopter*, National Advisory Committee for Aeronautics, Report No. 80, 1920.

the case of the cubic equation with complex coefficients should be derivable from the conditions given by P. Bohl[39] that the roots of a trinomial equation may have moduli less than p.

2.3. The case of equal roots. In the solution of linear differential equations with constant coefficients a double root of the associated algebraic equation indicates the existence of secular terms such as $a \sin (mt) + bt \cos (mt)$ or $(A + Bt)e^{-kt}$ in the general solution. This fact was a kind of bogie in the theory of the small oscillations of a dynamical system because it was thought at one time that there was a kind of instability associated with the presence of repeated roots. In 1858 Weierstrass[40] completed the theory of normal coordinates and showed that in some cases at least secular terms do not occur in the final solution of the equations of motion. Further remarks of interest were made by Somoff,[41] Routh[42] and Stokes.[43] A review of the subject has been given recently by Melikov.[44]

Experience shows that it is better to work with the original system of differential equations than with the single equation obtained by eliminating all the variables but one. For instance, in the case of the well known system

$$a(x'' + k^2 x) = b(y'' + k^2 y) = c(z'' + k^2 z) = bcx + cay + abz = s,$$

say, the algebraic equation for m in an exponential factor e^{mt} occurring in the solutions is $(m^2 + k^2)^2 (m^2 + h^2) = 0$ where $h^2 = k^2 - bc/a - ca/b - ab/c$. The equation for s is, however, $s'' + h^2 s = 0$ and so secular terms do not appear when s is calculated first and x, y, z derived from s.

[39] P. Bohl, *Zur Theorie der trinomischen Gleichungen*, Math. Ann. vol. 56 (1908) pp. 556–569.

[40] K. Weierstrass, *Über ein die homogenen Functionen zweiten Grades betreffendes Theorem, nebst Anwendung desselben auf die Theorie der kleinen Schwingungen*, Monatsberichte der Akademie der Wissenschaft zu Berlin, 1858, pp. 207–220; *Mathematische Werke*, vol. 1, Berlin, 1894, pp. 233–246. See also F. Purser, *Occurrences of equal roots in Lagrange's determinantal equation*, Report of the British Association for the Advancement of Science, 1878, pp. 463–464.

[41] K. Somoff, *Oscillations of systems of particles, algebraic problem*, Mémoires de l'Academie des Sciences de Saint Petersburg (Akademiia Nauk), no. 14, 1879, 30 pp. The related paper of C. Jordan, *Sur les oscillations infiniment petites des systèmes matériels*, C. R. Acad. Sci. Paris vol. 74 (1872) pp. 1395–1399, is reviewed unfavorably in Fortschritte der Mathematik vol. 4 (1872) pp. 471–472.

[42] E. J. Routh, *Rigid dynamics*, part 2, pp. 84, 190.

[43] G. G. Stokes, *Explanation of a dynamical paradox*, Messenger of Mathematics vol. 1 (1872) pp. 1–3; *Mathematical and physical papers*, vol. 4, pp. 334–335.

[44] K. V. Melikov, *Über das Theorem von Weierstrass und Routh*, Annals of the Institute of Mines, Leningrad, vol. 10 (1936) pp. 71–76.

The case of equal roots is connected of course with the phenomenon of resonance and there are many cases in practice in which a number of identical dynamical systems are coupled together particularly in the construction of acoustical, electrical and mechanical filters. Much depends on the nature of the coupling and even when equal roots do not occur in the final analysis there are interesting phenomena. The use of symmetrical arrangements is sometimes advantageous on account of the simplicity of the analysis.

In gas producing plants in which there is one exhaust regulator for two coke ovens there seems to be an idea that symmetry must be avoided on account of a possible interaction or resonance between two coke ovens which would make the regulator unstable. Thus Dougill[45] remarks: "The interaction which so often occurred when two retort houses of equal size were connected to a common main which led to one exhaust governor could be remedied by provision of a time lag, preferably in the exhaust governor."

The question may be raised whether the troubles encountered can really be attributed to the equality in size and an answer to this question cannot be given without a careful analysis of the precise setup. In the meantime, however, it may be of interest to examine some of the complications which arise when use is made of a time lag in dynamical or electrical systems.

3. THE TRANSCENDENTAL PROBLEM

3.1. **Time lag in control systems.** Long ago the delayed action of a regulating system was recognized as one of the primary causes of the hunting of governed engines.[46] The effect of time lag has consequently been studied by many investigators, particularly by D. R. Hartree, A. Porter, A. Callender, A. B. Stevenson,[47] H. König,[48] J. G. Ziegler

---

[45] G. Dougill, *Retort house and exhauster governing of gas works*, Engineering vol. 144 (1937) p. 144.

[46] See for instance, J. Swinburne, *The hunting of governed engines*, Engineering vol. 58 (1894) p. 247; Practice, *The "hunting" of steam engine governors*, Engineering vol. 71 (1901) p. 216.

[47] A. Callender, D. R. Hartree and A. Porter, *Time-lag in a control system*, Trans. Roy. Soc. London Ser. A. vol. 235 (1936) pp. 415–444. D. R. Hartree, A. Porter, A. Callender and A. B. Stevenson, *Time-lag in a control system. II*, Proc. Roy. Soc. London Ser. A. vol. 161 (1937) pp. 460–476. A. Callender and A. B. Stevenson, Proceedings of the Society of the Chemical Industry (Chemical Engineering Group) vol. 18 (1936) p. 108. See also L. Nisolle, *Sur la stabilité des régulateurs à impulsions retardies ou amorties*, C. R. Acad. Sci. Paris vol. 211 (1940) pp. 762–765.

[48] H. König, *Periodische und aperiodische Schwingungen an empfindlichen Regelan ordnungen*, Zeitschrift für Technische Physik vol. 18 (1937) pp. 426–431. See also D. Stein, *Untersuchung der Stabilitätsbedingungen bei verzögerter Regelung*, Elektrische Nachrichten Technik vol. 20 (1943) pp. 205–213.

and N. B. Nichols.[49] Time lag was also considered by H. L. Hazen[50] in his work on servo mechanisms and by N. Minorsky[51] in his study of control problems.

The effects of time lag have been considered usually by three distinct methods:

(1) By the use of Taylor's theorem and a neglect of small terms so that linear differential equations are obtained.

(2) By the use of differential difference equations or equations of mixed differences.

(3) By the use of integral equations of the Poisson-Volterra type.

The first method is explained in a general discussion of control problems by the editorial staff of The Engineer and by N. Minorsky who regards the differential equation as an asymptotic form and gives four different types. In a simple case a body is supposed to oscillate under the influence of a restoring force $R(t-k)$ proportional to the body's displacement at a previous instant and also under the influence of a damping depending partly on the body's instantaneous velocity and partly on its velocity at a previous time $t-h$. The equation of motion is supposed, indeed, to be of the type[52]

$$x'' + Qx' + Nf'(t - h) + Pf(t - k) = 0$$

where the time lags h and k are regarded as independent of t. In the approximate theory $f'(t-h)$ and $f(t-k)$ are replaced by $f'(t) - hf''(t)$ and $f(t) - kf'(t) + k^2f''(t)/2$ respectively and then x is used in place of $f(t)$. The resulting equation is

$$(1 - Nh + Pk^2/2)x'' + (Q + N - Pk)x' + Px = 0.$$

When $Q = N = 0$ the lag in R gives a negative damping and so oscillation with increasing amplitude may be expected to occur. When $N = 0$ and $Q \neq 0$ the negative damping may be overcome by positive damping depending on Q. When $N = 0$ both N and Q may tend to overcome the negative damping but if h is large the coefficient of x'' may become negative and completely alter the character of the motion. Minorsky[53] indicates an asymptotic form in which the apparent re-

[49] J. G. Ziegler and N. B. Nichols, *Process lags in automatic-control circuits*, Transactions of the American Society of Mechanical Engineers vol. 65 (1943) pp. 433–444.

[50] H. L. Hazen, *Servo mechanisms*, Journal of the Franklin Institute vol. 218 (1934) pp. 279–331, 543–580.

[51] N. Minorsky, *Control problems*, ibid. vol. 232 (1941) pp. 451–488.

[52] Editorial Staff, *The damping effect of time lag*, The Engineer vol. 163 (1937) p. 439.

[53] N. Minorsky, *Self-excited oscillations in dynamical systems possessing retarded actions*, Transactions of the American Society of Mechanical Engineers vol. 64 (1942) pp. A65–A71, discussion by H. Poritzky, pp. A195–A196.

storing term is different from Px, but by the multiplication of the equation by a suitable factor this form may be reduced to the previous ones in which only the coefficients of x' and x'' are affected by the lags.

The indications of the approximate theory need to be checked by exact analysis but they are such as to make it plain that the effects of time lag may be quite serious.

When there is a single time lag which is treated as constant some progress may be made with the aid of the known theory of linear equations of mixed differences. The particular equations discussed by Hartree and his collaborators are

$$u'(x) = f(x) + v(x) - cu(x),$$
$$- v'(x + 1) = pu(x) + qu'(x) + ru''(x)$$

where p, q, r and c are real constants. When $f(x) = 0$ the free motion is described by means of terms of type $u(x) = K \exp (kx)$ where k is determined by means of the transcendental equation

$$k(k + c) = e^{-k}(p + qk + rk^2).$$

This equation and some related equations are discussed chiefly by graphical methods but for the equation

$$z^2 e^z + a_0 z + a_1 = 0$$

an approximate solution $z = \log (a_0/b) + ib$, $b = (2n + 1/2)\pi + b^{-1}[\log (a_0/b) - a_1/a_0]$, is given for the value of z for a high harmonic on the supposition that b is large compared with a_1/a_0 and the real part of z. This approximation may hold in some cases for the fundamental and if it does it indicates that if the fundamental is positively damped the higher harmonics are more strongly damped.

Transcendental equations of the form $e^z = $ rational function of z are of frequent occurrence. A simple equation of this type $z = a - ce^{-z}$ occurs in economics[54] in the work of Kalecki, Frisch, Holme, James and Belz. It is a generalization of an equation considered by Euler[55] in 1750.

In his discussion of control problems König avoids the assumption

[54] M. Kalecki, *A macrodynamic theory of business cycles*, Econometrica vol. 3 (1935) pp. 327–344. R. Frisch and H. Holme, ibid. pp. 225–239. R. W. James and M. H. Belz, ibid. vol. 4 (1936) pp. 157–160.

[55] L. Euler, *Investigatio curvarum quae evolutae sui similes producunt*, Akademiia Nauk vol. 12 (1750) pp. 3–52. See also M. Alle, *Ein beitrag zur Theorie der Evoluten*, Akademie der Wissenschaften Wien (IIa) vol. 113 (1904) pp. 53–70.

of a constant time-lag and obtains an integral equation of Poisson's type,

$$x(t) + m \int_0^t k(t - T)x(T)dT = x_0(t),$$

where $x(t)$ is the variable quantity to be regulated, $x_0(t)$ is the variation of this quantity when there is no control, m is the factor of amplification and $k(t)$ is a function of type $I'(t)$, where $I(t)$ is the influence function which electrical engineers call the *transfer function*. The function $k(t)$ is generally zero up to time t_0, it then rises gradually in value until it reaches a peak value and then remains practically constant from a time t_1 on. The graph of $k(t)$ generally has a peak but in a simple case worked out by König, $k(t)$ is constant for $t_0 < t < t_1$ and is zero for other values of t. In a stable kind of regulation the free motion with $x_0(t) = 0$ is damped. When undamped oscillations or growing oscillations can arise the system may be capable of spontaneous oscillation.

König seeks the condition that there may be a solution of type $x(t) = A \exp(iwt - ht)$ and obtains the conditions

$$\int k(u)e^{hu} \sin(wu)du = 0, \qquad \int k(u)e^{hu} \cos(wu)du = 0.$$

The limit of stability is then given by $h = 0$. The motion due to a transitory disturbance may be found by Poisson's method of successive approximations[56] in which $x(t)$ is expanded in powers of m or it may be found by a method recommended by V. Pareto[57] and the present author[58] in which a relation is found between the generating functions

$$X(z) = \int_0^\infty e^{-zt}x(t)dt, \qquad K(z) = \int_0^\infty e^{-zt}k(t)dt,$$

$$X_0(z) = \int_0^\infty e^{-zt}x_0(t)dt.$$

[56] S. D. Poisson, *Mémoire sur la théorie du magnétisme en mouvement*, Académie des Sciences, Paris, 1826, 130 pp. (pp. 28–30).

[57] V. Pareto, *Sur les fonctions génératrices d'Abel*, J. Reine Angew. Math. vol. 110 (1892) pp. 29–323.

[58] H. Bateman, *Report on the history and present state of the theory of integral equations*, British Association for the Advancement of Science, 1910, pp. 345–424 (p. 394); *An integral equation occurring in a mathematical theory of retail trade*, Messenger of Mathematics vol. 49 (1920) pp. 1–4.

3.2. Feedback. An early use of feedback to regulate a water clock has been ascribed to James Watt[59] who apparently used a pump to maintain the desired constant level in the reservoir from which the water flows. The rate of flow should be constant if the level of the water in a receiving vessel is to give a correct measure of time. A cascade system of reservoirs which kept this rate very nearly constant for a short time was adopted long ago in the design of a water clock of Canton, China, known as "Hon-woo-et-low" (copper jars dropping water). James Arthur[60] saw this clock in 1897 and was told that it had been in existence for over 3000 years, being known as the clock of the street arch.

A mathematical theory of a cascade system of reservoirs based upon the formula for the discharge of a weir was given by E. Maillet[61] about 1905. The system of differential equations is nonlinear but some interesting conclusions are drawn relating to the existence of a steady state and the manner in which it is approached. The problem of stability of the steady state is considered and some attention is given also to the case in which water is fed into the reservoirs from an outside source. When in addition feedback is introduced there are many mathematical problems to be solved. Maillet's analysis is of some mathematical interest as it led him to researches on almost periodic functions.

Feedback has been much used in recent years in systems employing vacuum tubes and amplifiers. In his description of stabilized feedback amplifiers H. S. Black[62] says: "By building an amplifier whose gain is deliberately made, say, 40 decibels higher than necessary (10000 fold excess on energy basis) and then feeding the output back on the input in such a way as to throw away the excess gain, it has been found possible to effect extraordinary improvement in constancy of amplification and freedom from nonlinearity."

In the simplified mathematical theory which has been developed by

[59] See the discussion by Field of the paper by J. Woods, *Exhibition and description of the chronometric governor, invented by Messrs E. W. and C. W. Siemens*, Minutes and Proceedings of the Institute of Civil Engineers, London, vol. 5 (1846) pp. 255–265.

[60] James Arthur, *Time and its measurement*, Windsor, Chicago, 1909.

[61] E. Maillet, *Sur la vidage des systèmes de réservoirs*, C. R. Acad. Sci. Paris vol. 140 (1905) pp. 712–714; *Sur les équations différentielles et les systèmes de réservoirs*, ibid. vol. 147 (1908) pp. 966–968; *Sur les systèmes de réservoirs*, ibid. vol. 149 (1909) pp. 105–107. See also Bull. Soc. Math. France vol. 33 (1905) pp. 129–145; Annales des Ponts et Chaussées (1906) pp. 110–149; J. École Polytech. (2) vol. 13 (1909) pp. 27–56; J. Math. Pures Appl. (6) vol. 9 (1913) pp. 171–231.

[62] H. S. Black, *Stabilized feedback amplifiers*, Bell System Technical Journal vol. 13 (1934) pp. 1–18.

the electrical engineers[63] the effects of inertia or induction are neglected and the electrical system is supposed to be built up from simple
delay elements and elements with a constant type of amplification
over a limited range. The result of feedback is thus represented by
an equation

$$x = - Qx$$

where x denotes the amplification factor which is the product
$m_1 m_2 \cdots m_n$ of a number of individual amplification factors and Q is
the product of a number of differential operators of type

$$q = 1/(1 + \tau p), \qquad p = d/dt.$$

The differential equation for x is

$$[(1 + \tau_1 p)(1 + \tau_2 p) \cdots (1 + \tau_n p) + \mu]x = 0$$

and when a particular solution is of form $x = ae^{zt}$, z satisfies the algebraic equation obtained by replacing p by z. A graphical method of
finding the condition for stability has been given by H. Nyquist.[64]

It is clear from an algebraic standpoint that there is only one
condition because μ enters into only one of Hurwitz's determinants
and the others are automatically positive on account of the time constants τ. For a given set of time constants the system will generally
be stable when μ lies below a certain critical value μ_0 and unstable for
$\mu \geqq \mu_0$. When $\mu = \mu_0$ there can be one or more oscillations with constant amplitude. In the graphical form of the criterion there is stability when the point $(-1, 0)$ lies outside a certain curve traced out
by a radius vector representing the complex quantity μQ when p is
replaced by $i\omega$. The graphical method has been discussed by others.[65]
The differential equation for x can, of course, be replaced by an integral equation which is of the type considered by König or of a

[63] D. G. Prinz, *Contributions to the theory of automatic controllers and followers*,
Journal of Scientific Instruments vol. 21 (1944) pp. 53–64.

[64] H. Nyquist, *Regeneration theory*, Bell System Technical Journal vol. 11 (1932)
pp. 126–147; Annales des Postes, Télégraphes et Téléphones, Paris vol. 23 (1934)
pp. 1010–1016. See also K. Kupfmüller, *Über die Dynamik der selbsttägigen Verstärk-
ungsregler*, Elektrische Nachrichten Technik vol. 5 (1928) pp. 459–467.

[65] R. Feiss, *Bestimmung der Regelungsstäbilitat an Hand des Vektorbildes*, Zeitschrift
für der Verein Deutsches Ingenieures vol. 84 (1940) pp. 819–824. E. Peterson,
J. G. Kreer and L. A. Ware, *Regeneration theory and experiments*, Proceedings of the
Institute of Radio Engineers vol. 22 (1934) pp. 1191–1210, Bell System Technical
Journal vol. 13 (1934) pp. 680–700. D. G. Reid, *Necessary conditions for stability (or
self oscillation) of electrical circuits*, Wireless Engineers vol. 14 (1937) pp. 588–596.
C. A. A. Wass, *Feedback amplifiers*, Nature vol. 150 (1942) pp. 381–382.

slightly more general type. In this connection it may be worth while to recall the investigations of P. Hertz[66] and G. Herglotz[67] on natural vibrations of an electron. The integral equation considered was then of the form

$$x(t) = \int_0^\gamma x(t - T)q(T)dT$$

and the problem was to find the complex roots of the equation

$$\int_0^\gamma e^{-pT}q(T)dT = 1.$$

The foregoing theory of feedback is based on linear differential equations with constant coefficients and is only approximate. Actually the resistances and capacities may vary with frequency and may even vary with time. In radio-telephony the voice acts so as to modify the resistance of the oscillatory circuit or the capacity of its condenser. J. R. Carson[68] proposed a differential equation with periodic coefficients as a basis of a theory of modulation and the theory has been worked out more fully by O. Emersleben,[69] W. L. Barrow,[70] and A. Erdélyi.[71] Conditions of stability are obtained with the aid of the theory of integral equations and of asymptotic solutions of linear differential equations. Feedback is not always desirable. In a discussion of receivers and transmitters for demonstrating frequency modulation M. Hobbs[72] says that in order to avoid acoustical feedback it is necessary to locate the signal generator and microphone in one studio and the receivers in another.

When the differential equations of the system are nonlinear the theory of stability or of sustained oscillations is more difficult but

---

[66] P. Hertz, *Die Bewegung eines Elektrons unter dem Einflusse einer stets gerichteten Kraft*, Math. Ann. vol. 56 (1908) pp. 1–86.

[67] G. Herglotz, *Über die Integralgleichungen der Elektronentheorie*, ibid. pp. 87–106.

[68] J. R. Carson, *Notes on the theory of modulation*, Proceedings of the Institute of Radio Engineers vol. 10 (1922) pp. 57–64.

[69] O. Emersleben, *Natural oscillation of circuits containing variable capacities and resistances*, Physikalische Zeitschrift vol. 22 (1921) pp. 393–400.

[70] W. L. Barrow, *Frequency modulation and the effects of a periodic capacity variation in a non-dissipative oscillatory circuit*, Proceedings of the Institute of Radio Engineers vol. 21 (1933) pp. 1182–1202.

[71] A. Erdélyi, *Über die freien Schwingungen in Kondensatorkreisen mit periodisch veränderlicher Kapazität*, Annalen der Physik (5) vol. 19 (1934) pp. 585–622.

[72] M. Hobbs, *A low-power transmitter for demonstrating F-M receivers*, Electronics vol. 14 (1941) pp. 20–23.

there is a large literature on the subject.[73] Nonlinear feedback oscillations have been discussed by G. Hakata and M. Abe.[74]

3.3. Transcendental equations in the theory of integral equations.
With the abbreviations

$$(xgy) = \int_0^1 \int_0^1 x(s)g(s,\,t)y(t)dsdt, \qquad (uv) = \int_0^1 u(s)v(s)$$

a brief study will be made of the linear integral equations

$$(1) \quad f(s) = \int_0^1 g(s,\,t)F(t)dt + \lambda \int_0^1 h(s,\,t)F(t)dt + \lambda^2 \int_0^1 k(s,\,t)F(t)dt,$$

$$(2) \quad f(s) = F(s) + \lambda \int_0^1 h(s,\,t)F(t)dt + \lambda^2 \int_0^1 k(s,\,t)F(t)dt.$$

When λ is a complex quantity $a+ib$ where a and b are real and $f(s)$
is regarded as independent of λ and real, the solution $F(t)$ will also
be a complex quantity $u(t)+iv(t)$ with $u(t)$, $v(t)$ real provided the
kernels $g(s,\,t)$, $h(s,\,t)$, $k(s,\,t)$ are real for real values of s and t which
lie between 0 and 1. The combination $u(t)-iv(t)$ will be denoted by
the symbol $F^*(t)$ and for both equations the properties of the function

$$w(\lambda) = \int_0^1 f(s)F(s)ds, \qquad w(\lambda^*) = \int_0^1 f(s)F^*(s)ds$$

will be studied. If c is a real constant the zeros and poles of the function $w(\lambda)-c$ will be pseudo-negative when the same is true for the
zeros and poles of the function $w(\lambda^*)-c$.

In the important case in which $g(s,\,t)=g(t,\,s)$, $h(s,\,t)=h(t,\,s)$,
$k(s,\,t)=k(t,\,s)$ it is readily seen that in the two cases

$$(1') \quad \begin{aligned} w(\lambda^*) &= (ugu)+(vgv)+a[(uhu)+(vhv)]+(a^2-b^2)[(uku)+(vkv)] \\ &\quad +ib[(uhu)+(vhv)]+2iab[(uku)+(vkv)], \end{aligned}$$

$$(2') \quad \begin{aligned} w(\lambda^*) &= (uu)+(vv)+a[(uhu)+(vhv)]+(a^2-b^2)[(uku)+(vkv)] \\ &\quad +ib[(uhu)+(vhv)]+2iab[(uku)+(vkv)]. \end{aligned}$$

The right-hand sides of these equations are zero when λ is such that
$w(\lambda^*)=0$ and also when λ is such that $F(t)$ exists when $f(s)=0$. In
the important case in which the functions h and k are of positive type

[73] K. Heegner, *The self-oscillating vacuum tube*, Arkiv för Elektrot. vol. 9 (1920)
pp. 127–152.

[74] G. Hakata and M. Abe, *Non-linear differential feedback oscillations*, Nippon
Electrical Communication Engineering no. 5 (1939) pp. 526–536.

the integrals (uhu), (vhv), (uku), (vkv) are all positive and so when the imaginary terms on the right are equated to zero it is seen that either a is negative or b is zero. When b is zero the equation obtained by equating the real part of the right-hand side to zero indicates that a is negative in case $(2')$ and this is true also in case $(1')$ if $g(s, t)$ is also of positive type. If c is negative it is readily seen that a must be negative when $w(\lambda^*) = c$.

When $k(s, t) = 0$ and $g(s, t)$, $h(s, t)$ are of positive type it is known that the zeros and poles of the function $w(\lambda)$ are all negative and occur alternately. The situation is analogous to that which occurs in the theorem of Routh, Jouguet and Chipart relating to the even and odd parts of an algebraic equation with pseudo-negative roots and so the function $w(\lambda)$ can be used quite often to construct a transcendental equation with only pseudo-negative roots. As an example of the first theorem we take equation (2) with

$$h(s, t) = pst, \quad k(s, t) = s(1 - t) \quad \text{or} \quad t(1 - s)$$

according as $s \lessgtr t$; the equation for λ is then

$$\coth \lambda = (1/\lambda) - (1/p)$$

when λ is a pole. When the equation is written in the form

$$\text{ch } (\lambda) - (1/\lambda) \text{ sh } (\lambda) + (1/p) \text{ sh } (\lambda) = E(\lambda) + O(\lambda) = 0$$

it is seen that

$$E(z^{1/2}) = \text{ch } (z^{1/2}) - z^{-1/2} \text{ sh } (z^{1/2}), \qquad z^{-1/2}O(z^{1/2}) = z^{-1/2} \text{ sh } (z^{1/2}).$$

It is readily seen that the functions on the right are transcendental functions of z with negative zeros which occur alternately.

In the second theorem if $h(s, t) = s(1 - t)$ or $t(1 - s)$ according as $s \lessgtr t$ it is found that if

$$C(z) = \int_0^1 \text{ch } (zt)f(t)dt, \qquad S(z) = \int_0^1 \text{sh } (zt)f(t)dt$$

then the equation

$$0 = \text{sh } (z) + z \text{ sh } (z)(ff) - z \text{ sh}^2 (z)S(z)C(z) + \text{sh } (z) \text{ ch } (z) [S(z)]^2$$

has only pseudo-negative roots.

4. THE SEPARATION OF VIBRATIONS

4.1. Acoustical filters. The early work of Poisson on the propagation of sound along a branched pipe was followed by the inventions of John Herschel and Quincke for the production of interference of waves by the rejunction of the divided branches of a pipe. The theory based on the idea of velocity potential and simplified boundary conditions was improved by Stewart and others and then replaced by a theory of lumped impedances so that the theory of acoustic filters could be developed along the same lines as the theories of mechanical and electrical filters. A good account of the theory from this standpoint is given in the book of Stewart and Lindsay.

The filtering action of a regularly spaced series of similar sheets of muslin was considered by Rayleigh[75] in the period 1887-1896 and is described in a passage inserted in the 1896 edition of his *Theory of sound*. He states that if a moderate number of such sheets be placed parallel to one another and at such distances apart that the partial reflections agree in phase, then a sensitive flame may be powerfully affected. With the aid of a device for adjusting the interval between two consecutive sheets it is easy to find how this interval depends on the wave length λ when the condition for effective reflexion is satisfied. Rayleigh states that with $a = \lambda/2$ the condition is satisfied for normal incidence but in the actual experiment it is more convenient to use oblique incidence and the calculations necessary for this case are readily made.

In his mathematical investigations[76] Rayleigh considered the transverse vibrations of a stretched string periodically loaded, but the analysis is rather difficult as it depends on the properties of the solutions of differential equations with periodic coefficients and use is made of infinite determinants as in the work of G. W. Hill. The vibrations of this type of string in which the density varies continuously have been studied further by Strutt[77] but more progress has been made in the study of the older problem in which the density of the string varies discontinuously. This case will be considered later for both transverse and longitudinal vibrations. The former case is interesting on account of analogies with optical phenomena, the latter on account of the analogies with acoustical phenomena.

The problem of the loaded string was much studied by the great

[75] Lord Rayleigh, *Iridescent crystals*, Proceedings of the Royal Institute of London vol. 12 (1889) pp. 447-449; Nature vol. 40 (1889) pp. 227-228; *Scientific papers*, vol. 3, pp. 264-266 (see also pp. 1-14, 204-212); *Theory of sound*, vol. 2, p. 311.

[76] Lord Rayleigh, *On the maintenance of vibrations by forces of double frequency, and on the propagation of waves through a medium endowed with a periodic structure*, Philosophical Magazine vol. 24 (1887) pp. 145-159; *On the remarkable phenomenon of crystalline reflexion described by Stokes*, ibid. vol. 26 (1888) pp. 256-265.

[77] M. J. O. Strutt, *Eigenschwingungen einer Saite mit sinusformiger Massenverteilung*, Annalen der Physik vol. 85 (1928) pp. 129-136.

mathematicians of the eighteenth century and many of their results are given by Routh in his *Rigid dynamics*. Routh also mentions that in April 1875 Lord Kelvin studied the vibrations and waves in a stretched uniform chain of symmetrical gyrostats connected together by universal flexure joints. His thoughts returned to this subject in his Baltimore lectures of 1884.

The theory of the loaded string became definitely associated with the theory of mechanical filters in 1898 when Godfrey and Lamb published their researches and when Campbell and Pupin became interested in the properties of the loaded electrical transmission line. Acoustic filters are much used as mufflers for internal combustion engines and as may be seen from the lists of patents in the Journal of the Acoustical Society of America baffles, holes and side branches in the exhaust pipe are among the devices used. Resonating side branches sometimes communicate with one another through partitions of absorbing material. Even a long pipe has a filtering action as the attentuation is higher for sound of some frequencies than for sound of some other ranges of frequency. Problems relating to pipes will be discussed in the section dealing with hydrodynamics in which some consideration will be devoted not only to the elimination of noise but also the reduction of dangerous vibrations in hydraulic pipe lines.

4.2. Passage of sound through a slab. Let ρ, v, and $Z = \rho v$ be the density velocity of sound and radiation resistance of a homogeneous slab of thickness a which is of infinite extent in any direction parallel to the plane faces. Let ρ, v', $Z' = \rho' v'$ be the corresponding quantities for the medium outside the slab. For normal incidence of waves on the face $x = 0$ the velocity potentials are

$$\phi = De^{iw(t-x/v')} + Ee^{iw(t+x/v')} \qquad \text{for } x \leqq 0,$$

$$\phi = Be^{iw(t-x/v)} + Ce^{iw(t+x/v)} \qquad \text{for } 0 \leqq x \leqq a,$$

$$\phi = Ae^{iw(t-x/v')} \qquad \text{for } x \geqq a.$$

The boundary conditions are

$$\rho'(D + E) = \rho(B + C), \qquad \rho(Be^{-is} + Ce^{is}) = \rho'Ae^{-is'},$$

$$is'(E - D) = is(C - B), \qquad is(Ce^{is} - Be^{-is}) = -is'Ae^{-is'}$$

where $s = wa/v$, $s' = wa/v'$. Thus

$$A/D = 2e^{is}/[2 \cos (s) + i(Z/Z' + Z'/Z) \sin (s)]$$

and the coefficient of reflection is

$$r = (Z'/Z - Z/Z')[4 \cot^2 (s) + (Z'/Z - Z/Z')^2]^{-1/2}.$$

This is the formula of Lord Rayleigh.[78] Interesting applications of this formula to the reflection and transmission of sound through partitions have been made by Boyle[69] and Davis.[80] It is clear that $r = 0$ when sin $(s) = 0$ and so there are certain critical thicknesses a for which there is no reflection of sound waves of the prescribed frequency $w/2\pi$. Davis regards the formula as applicable to the transmission of sound through light thin panels such as sheets of paper, sailcloth or fibre board. When a is very small there is a tendency for the reduction factor of the energy to vary as the square of the frequency f of the incident sound. With materials as light as paper a term due to air damping is important and there is less variation with f. For heavy panels such as two inch boards or brick walls the reduction factor is less than that given by Rayleigh's formula. Davis has given a formula

$$\text{Reduction factor} = (1/2R)^2[(r + 2R)^2 + (m/w)^2(w^2 - w_0^2)]$$

which indicates that resonances can account for a reduced insulating value. This formula is derived from a differential equation of type

$$m\ddot{\xi} + (r + 2R)\dot{\xi} + S\xi = 2R\xi_0 e^{iwt}, \qquad w_0 = (S/m)^{1/2}.$$

To account for the behavior of actual panels it seems necessary to assume that there are several modes of vibration with which there can be resonance. This is in accordance with the general theory of the vibration of plates and with experiment as is pointed out by Davis and Littler.[81]

In the work of Boyle on the influence of the thickness of the plate on the transmission of sound, use was made of a high $f(135000\sim$, $300000\sim$ and $528000\sim$). It was found that when a was a large multiple of $\lambda/4$ (where λ is the wave length in the plate) there was a maximum of energy reflected and a minimum of energy transmitted. When a was a few integral multiples of $\lambda/2$ there was an almost complete transmission of energy. Thus in the latter case the plate acted as a

---

[78] Lord Rayleigh, *Theory of sound*, vol. 2, 1896, p. 88.

[79] R. W. Boyle, *Transmission of sonic and ultrasonic waves through partitions*, Nature vol. 121 (1928) pp. 55–56. See also R. W. Boyle and D. K. Froman, Canadian Journal of Research vol. 1 (1929) pp. 405–424.

[80] A. H. Davis, *Transmission of sound through partitions*, Philospohical Magazine (7) vol. 15 (1933) pp. 309–316.

[81] A. H. Davis and T. S. Littler, *The measurement of transmission of sound by partitions of various materials*, Philosophical Magazine (7) vol. 3 (1927) pp. 177–194, vol. 7 (1929) pp. 1050–1062.

high pass filter and in the former case as a high frequency rejector. This work has been continued by Boyle and Sproule[82] with the aid of a torsion pendulum and the previous conclusion confirmed.

4.3. **Multiple partitions.** The passage of sound through several slabs of different materials is of some interest but the analysis is rather complicated. The case of three different media was considered by Brillié and his work is discussed by Stewart and Lindsay. The case of five media two pairs of which are alike in properties and are symmetrically related to a third medium in the middle is of much interest in relation to the sandwich type of radiator and receiver which was used at one time in underwater work. The natural vibrations of such a sandwich are of interest.

The vibrations of a column of gas, one portion of which is at a temperature T_1 and the other at a temperature T_2 have been studied by Lees[83] in connection with some experiments on the vibration of travelling flames made by Coward and Hartwell.[84] Lees considered 3 cases: $1°$. Tube closed at both ends. $2°$. Tube open at both ends. $3°$. Tube closed at one end and open at the other. In the last named case the frequency of vibration is determined by the equation

$$(n_1/F_1) \tan (m_1\pi/2) = - (n_2/F_2) \tan [(m_2 - 1)(\pi/2)]$$

where n_1 is the fundamental frequency when the whole column is at temperature T_1, n_2 is the corresponding frequency when the temperature is T_2, F_1 and F_2 are the moduli of adiabatic elasticity for longitudinal displacements in the column at the two temperatures and

$$m_1 = na/n_1c, \qquad m_2 = nb/n_2c, \qquad a + b = c$$

where a, b are the lengths of the two portions of the tube.

Meyer observes that it is well known that the sound-damping action of a homogeneous wall increases with its thickness provided the exciting frequency is sufficiently higher than the natural frequency of vibration of the wall—a condition that is usually satisfied. When several such walls have air between them the composite wall acts as a mechanical damper because the layers of air form buffers if they are small compared with the wave length. If m denotes the mass of the

[82] R. W. Boyle and D. O. Sproule, *Transmission of sound energy and thickness of plate transmitter at normal incidence-ultrasonic method*, Canadian Journal of Research vol. 2 (1930) pp. 3–12.

[83] C. H. Lees, *Free periods of a composite elastic column or composite stretched wire*, Proceedings of the Physical Society of London vol. 41 (1929) pp. 204–213.

[84] J. F. Coward and F. J. Hartwell, *Extinction of methane flames by diluent gases*, Journal of the Chemical Society vol. 129 (1926) pp. 1522–1532.

wall in grams per square centimeter, if l is the length of an air buffer, if v is the velocity of sound and if ρ is the density of the air, the fundamental natural frequency is

$$f_0 = (v/\pi)(\rho l/m)^{1/2}.$$

This formula has been confirmed over a frequency range from 400$\sim$ to 7000$\sim$. Meyer[85] states that in such composite walls the velocity of transmission decreases as the frequency increases but the damping does not seem to increase very rapidly. Investigations led to the following rules for the design of composite walls. The dimensions should be such that f_0 is less than the practically important range of f's and the cross vibrations in the air buffers should be damped. By following these rules sound can be effectively damped without the use of heavy walls. A four-fold composite wall forty centimeters thick (15 3/4 inches) weighing 50 kilograms per square meter (10.25 pounds per square foot) gave better insulation than a solid brick wall weighing 1000 kg. per sq.m. The effect of composite walls has been discussed by many other writers[86] and the mathematical theory has been elucidated by Constable.[87] For $f > f_0$ insulation is at first decreased as the separation of the walls is increased, but it afterwards increases continuously up to a point at which the separation is approximately $\lambda/4$. After this point it decreases to a second minimum and thereafter minima occur at successive increases of $\lambda/2$ in the separation of the walls. The first minimum at which the insulation can be less than that of one component alone can be attributed to the effect of air coupling.

The properties of a double partition constructed from dissimilar components were examined by Renault[88] and by Constable.[87] The latter found that at f's for which the resonances of the components

[85] E. Meyer, *Die Mehrfachwand als akustisch-mechanische Drosselkette*, Zeitschrift für Technische Physik vol. 16 (1936) pp. 565–566; *Über das Schallschluckvermögen schwingungsfähiger, nichtporöser Stoffe*, Elektrische Nachrichten Technik vol. 13 (1936) pp. 95–102.

[86] E. Wintergast, *Theorie der Schalldurchlassigkeit von einfachen und zusammengesetzten Wänden*, Schalltechnik vol. 4 (1931) pp. 85–91, vol. 5 (1932) pp. 1–8. J. E. R. Constable and G. H. Aston, *The sound insulation of single and complex partitions*, Philosophical Magazine (7) vol. 23 (1937) pp. 161–181. E. Lubcke and A. Eisenberg, *Zur Schallübertragung von dünnen Einfachwänden*, Zeitschrift Technische Physik vol. 18 (1937) pp. 170–174.

[87] J. E. R. Constable, *The acoustical insulation afforded by double partition constructed from similar components*, Philosophical Magazine (7) vol. 18 (1934) pp. 321–343; *Acoustical insulation afforded by double partitions constructed from dissimilar components*, ibid. vol. 26 (1938) pp. 253–259.

[88] L. Renault, *La transmission du son à travers cloisons métalliques*, Revue d'Acoustique vol. 6 (1937) pp. 69–101.

and the air coupling resonance can be neglected, the insulation to be obtained from a double partition for a given total weight and thickness is greatest when the components are similar.

The transmission of sound by a series of equidistant similar partitions has been studied by Hurst[89] whose results are very similar to those of Lamb in his first example of a mechanical filter (§4.5). The method of indicating the regions of attenuation is also similar to that of Lamb. Hurst gives also a theory of transmission through a series of circular panels each of which is set into a rigid wall. In this theory effects of diffraction are taken into account.

4.4. Results found by energy methods. The absorption of sound by a porous wall was studied theoretically by Rayleigh with the aid of a theory of the propagation of sound in a capillary tube which will be examined in §6.[89a] There are, however, some results which can be obtained by energy methods which may be mentioned here for comparison with the other results.

4.4a. The passage of energy through a single absorbing wall. By means of an extension of the analysis of Buckingham and Eckhardt,[90] Davis[91] has obtained the following equations for two rooms separated by an absorbing wall:

$$4V\dot{I} + vaSI = 4E + vkWI, \qquad 4V_1\dot{I}_1 + va_1S_1I_1 = vkWI.$$

The first equation refers to the room containing the source of sound which is supposed to emit energy at a constant rate E while operative. W denotes the area of the wall used as a partition between the two rooms. V, V_1 are the volumes of the rooms; S, S_1 are the respective total areas of the exposed surfaces of the walls of these rooms (including the partition); a, a_1 are the mean fractions of incident energy lost by the respective rooms at each reflection by absorption or transmission to other rooms; v is the velocity of sound and k is a factor of type a for the partition. In the steady state $\dot{I}=\dot{I}_1=0$ so the maximum values of I and I_1 are J and J_1 respectively, where

[89] D. G. Hurst, *The transmission of sound by a series of equidistant partitions*, Canadian Journal of Research vol. 12 (1935) pp. 398–407.

[89a] References to §§5 and 6 refer to material which the author hopes will appear after the war.

[90] E. A. Eckhardt, *The acoustics of rooms, reverberation*, Journal of the Franklin Institute vol. 195 (1923) pp. 799–814.

[91] A. H. Davis, *Reverberation equation for two adjacent rooms connected by an incompletely sound-proof partition*, Philosophical Magazine (6) vol. 50 (1925) pp. 75–80. See also C. F. Eyring, *Methods of calculating the average coefficient of sound absorption*, Journal of the Acoustical Society vol. 4 (1933) pp. 178–192.

$$J = \frac{4Ea_1S_1}{v(aSa_1S_1 - k^2W^2)}, \qquad J_1 = JkW/(a_1S_1).$$

At time t after the source has been cut off

$$\frac{I}{J} = \frac{\lambda_2 - \lambda_0}{\lambda_2 - \lambda_1} e^{-\lambda_1 t} + \frac{\lambda_1 - \lambda_0}{\lambda_1 - \lambda_2} e^{-\lambda_2 t},$$

$$\frac{I_1}{J_1} = \frac{\lambda_2}{\lambda_2 - \lambda_1} e^{-\lambda_1 t} + \frac{\lambda_1}{\lambda_1 - \lambda_2} e^{-\lambda_2 t}$$

where

$$\left.\begin{array}{c} \lambda_1 \\ \lambda_2 \end{array}\right\} = (v/8)\{N + N_1 \mp [(N_1 - N)^2 + 4k^2W^2/VV_1]^{1/2}\},$$

$$N = aS/V, \qquad N_1 = a_1S_1/V_1, \qquad \lambda_0 = 4V_1\lambda_1\lambda_2/(va_1S_1).$$

4.4b. **The passage of waves through a series of interceptors.** The reflection of light from a pile of plates was discussed in a restricted form by Augustin Fresnel[92] and Franz Neumann, the formulae of the latter being quoted by Wild in 1856. In 1862 Stokes[93] treated the problem in a general way by means of the functional equations

$$r(m + n) = r(m) + r(n)[t(m)]^2/[1 - r(m)r(n)],$$

$$t(m + n) = t(m)t(n)/[1 - r(m)r(n)]$$

in which $r(m)$ denotes the fraction of energy reflected from a pile of m plates, $t(m)$ denotes the fraction of energy transmitted and $a(m) = 1 - r(m) - t(m)$ is the fraction of energy absorbed in the plates. In the derivation of these equations the plates are supposed to be all formed of the same material and to be all of the same thickness. The plates themselves and the interposed sheets of air are supposed to be so thick that the phenomenon of the colours of thin plates does not occur to any appreciable extent. The analysis deals only with intensities of light or other wave motion.

If $r(1) = r$, $t(1) = t$ and quantities a, b are defined by the equations

$$r \operatorname{ch} a + t \operatorname{ch} b = 1, \qquad r \operatorname{sh} a - t \operatorname{sh} b = 0,$$

[92] A. Fresnel, *Calculs sur les intensités de lumière réfléchi par une, deux et quatre glacés*, Oeuvres complètes, vol. 2, pp. 789–792.

[93] Sir George Stokes, *On the intensity of the light reflected from or transmitted through a pile of plates*, Proc. Roy. Soc. London vol. 11 (1862) pp. 545–556. J. Stirling, *Note on a functional equation treated by Sir George Stokes*, Proc. Roy. Soc. London Ser. A vol. 90 (1914) pp. 237–239.

an appropriate solution is

$$r(n) = \frac{\text{sh }(nb)}{\text{sh }(a + nb)}, \qquad t(n) = \frac{\text{sh }(a)}{\text{sh }(a + nb)}.$$

It should be noticed that

$$4 \text{ sh}^2 (a/2) = (r^{-1/2} - r^{1/2})^2 - t^2/r, \quad 4 \text{ sh}^2 (b/2) = (t^{-1/2} - t^2)^2 - r^2/t.$$

Since $1 - r > t$ and $1 - t > r$ the quantities a and b defined by these equations are both real. It should be noticed also that

$$t^2 = 1 - 2r \text{ ch } a + r^2, \qquad r^2 = 1 - 2t \text{ ch } b + t^2,$$
$$[t(n)]^2 = 1 - 2r(n) \text{ ch } a + [r(n)]^2,$$
$$[r(n)]^2 = 1 - 2t(n) \text{ ch } (nb) + [t(n)]^2.$$

Hence in order to investigate the effect of frequency on $r(n)$ and $t(n)$ it is useful to have plots in the rt-plane of the two systems of hyperbolas $a = \text{constant}$, $b = \text{constant}$.

4.5. **Mechanical filters.** A general theory of mechanical filters was given by Horace Lamb[94] and Charles Godfrey[95] in 1898. In the analysis of Lamb dynamical systems of any degree of complexity but all exactly alike are supposed to be interpolated at regular intervals along a line which is regarded for simplicity as an infinite string capable of longitudinal vibrations. The position and configuration of any one of these systems is imagined to be determined by the coordinate ξ of the point of the string where it is attached and by means of n other coordinates q_s $(s = 1, 2, \cdots, n)$. The kinetic energy T and potential energy V are, moreover, assumed to have the forms

$$2T = \sum_{r=1}^{n} a_r \dot{q}_r^2 + 2 \sum_{r=1}^{n} \alpha_r \dot{\xi} \dot{q}_r + P \dot{\xi}^2, \quad 2V = \sum_{r=1}^{n} b_r q_r^2 + 2 \sum_{r=1}^{n} \beta \xi q_r + Q \xi^2,$$

so that the equations of motion are

$$a_r \ddot{q}_r + b_r q_r + \alpha_r \ddot{\xi} + \beta_r \xi = 0, \qquad r = 1, 2, \cdots, n,$$

$$P \ddot{\xi} + Q \xi + \sum_{r=1}^{n} (\alpha_r \ddot{q}_r + \beta_r q_r) = X$$

where X is the external force corresponding to the coordinate x repre-

---

[94] H. Lamb, *On waves in a medium having a periodic discontinuity of structure*, Memoirs of the Manchester Literary and Philosophical Society vol. 42, no. 3, 1898, 20 pp.

[95] C. Godfrey, *Discontinuities of wave-motion along a periodically loaded string*, Philosophical Magazine (5) vol. 45 (1898) pp. 356–363.

senting in Lamb's model the difference of tensions on the two sides of the gap into which the dynamical system is introduced.

When all quantities contain the time factor exp $(ikct)$ the first n equations of motion give the relations

$$q_r = -\frac{k^2c^2\alpha_r - \beta_r}{k^2c^2a_r - b_r}, \qquad r = 1, 2, \cdots, n,$$

and the last equation gives

$$\left\{ k^2c^2P - Q - \sum_{r=1}^{n} (k^2c^2\alpha_r - \beta_r)^2 / (k^2c^2a_r - b_r) \right\} \xi = -X.$$

The successive dynamical systems may be distinguished by suffixes (D_r) and may be pictured as particles at intervals of length a. If $2\pi/k$ is the wave length for a disturbance with the same time factor on the unloaded string, the motion to the right of D_r is of type

$$\xi = \xi_r \cos (kx) + \left[\xi_{r+1} - \xi_r \cos (ka) \right] \sin (kx) \csc (ka)$$

while on the left it is

$$\xi = \xi_r \cos (kx) + \left[\xi_r \cos (ka) - \xi_{r-1} \right] \sin (kx) \csc (ka).$$

The tensions of the string on the two sides of D_r are

$$E(d\xi/dx)_+ = kE \csc (ka) \left[\xi_{r+1} - \xi_r \cos (ka) \right],$$
$$E(d\xi/dx)_- = kE \csc (ka) \left[\xi_r \cos (ka) - \xi_{r-1} \right],$$

so

$$X_r = kE \csc (ka) \left[\xi_{r+1} - 2\xi_r \cos (ka) + \xi_{r-1} \right]$$

and there is a difference equation

$$\xi_{r+1} - 2\left[\cos (ka) - f(k) \sin (ka) \right] \xi_r + \xi_{r-1} = 0$$

where

$$2kE \cdot f(k) = k^2c^2P - Q - \sum_{r=1}^{n} \frac{(k^2c^2\alpha_r - \beta_r)^2}{k^2c^2a_r - b_r} \, .$$

The solution of the difference equation has different forms according as the coefficient of $2\xi_r$ does or does not lie between the limits ± 1. The critical values of k are determined by

$$C(ka) \equiv \cos (ka) - f(k) \sin (ka) = \pm 1.$$

The roots of this equation give the ranges of frequency within which there is total reflection or partial transmission.

When $C(ka)$ lies between ± 1 there is a real angle θ such that

$$\cos \theta = C(ak)$$

and then the solution of the difference equation is

$$\xi_r = Pe^{ir\theta} + Qe^{-ir\theta}$$

where P and Q are suitable constants independent of r. The wavelength λ in the loaded medium may be defined to be

$$\lambda = (2\pi a/\theta) = (ka/\theta)\lambda_0$$

where $\lambda_0 = (2\pi/k)$ is the wave length in the unloaded medium. The wave velocity in the loaded medium may also be regarded as given by the equation

$$v = (\lambda kc/2\pi).$$

The group velocity is then

$$w = v - \lambda dv/d\lambda = -(\lambda^2 c/2\pi a)[d(ka)/d\theta](d\theta/d\lambda) = -[c/C'(ak)] \sin \theta.$$

Unless $C'(ak) = 0$ for the critical values of k the group velocity will be zero when k has a critical value and $\sin \theta = 0$. When θ lies between 0 and π the group velocity is positive when $C'(ak)$ is negative. The behavior of v may be found by Lamb's graphical method in which use is made of the curves

$$y = \cot (x/2), \qquad y = -\tan (x/2), \qquad y = f(x/a),$$

and the points of intersection of the last curve with the first two. With many forms of the function f the intervals of partial transmission are represented by intervals beginning respectively at $\pi, 2\pi, 3\pi, \cdots$, and the intervals of total reflection by intervals ending at these points. There may also be an interval of partial transmission between 0 and π.

In Lamb's first example $f(k) = \mu ka/2$, where $\mu = M/\rho a$. The dynamical system then consists of a mass M and ρa is the mass of the portion of the string between two consecutive masses. When the string is infinite in both directions there is transmission only for certain values of k below a certain limit. When ka is large θ is given approximately by $\theta^2 = k^2 a^2(1+\mu)$ and the refractive index N is

$$N = (\lambda/\lambda_0) = (1 + \mu)^{1/2}.$$

The effect of the loads is to increase the average density of the medium.

In Lamb's second example the mass M is urged towards its mean position by a spring and so

$$f(k) = \frac{\mu}{2}\left(ka - \frac{\sigma^2 a^2}{c^2 ka}\right).$$

The chief difference between this case and the last is that there is an interval of total reflection beginning at $ka = 0$ which is separated by an interval of partial transmission from the interval of total reflection which ends at $ka = \pi$. Lamb regards this as remarkable.

In Lamb's third example $f(k) = (\mu ka/2)(1 - k^2 c^2/\sigma^2)$. The chief difference between this case and the first is that as the frequency increases the intervals of partial transmission become wider and wider instead of shorter and shorter so that the medium is transparent for short waves of high frequency. This example was offered by Lamb to illustrate the transparency of a medium to Röntgen rays, a phenomenon which Stokes[96] had endeavoured to explain in his Wilde lecture of 1897.

In Lamb's model there are a number of equal light rigid circular frames each of which is attached to the string at opposite ends of a diameter and has in its interior a particle M connected with the frame by means of similar springs so that the particle is at the center of the frame when the springs are equally extended. If ξ_s denotes the displacement of the point of the string to which one of the frames is attached and if η_s denotes the displacement of the associated particle, the equation of motion of this particle is

$$M\ddot{\eta}_s + M\sigma^2(\eta_s - \xi_s) = 0$$

while there is a balance of tensions if

$$0 = kE[\xi_{s+1} - \xi_s \cos(ka)] - kE[\xi_s \cos(ka) - \xi_{s-1}]$$
$$+ M\sigma^2(\eta_s - \xi_s) \csc(ka).$$

These equations give the relation between ξ_{s+1}, ξ_s and ξ_{s-1} of the type mentioned.

Lamb's model furnishes a particular type of high pass filter.

4.6. **Selective reflection.** In Lamb's work the string is unloaded to the left of 0 and on the right has masses M at the points $x = 0, a, \cdots, na$ and is unloaded beyond the point na. For $x < 0$ it is assumed that

$$\xi = e^{ik(ct-x)} + Ae^{ik(ct+x)}.$$

---

[96] Sir George Stokes, *On the nature of the Röntgen rays*, Memoirs of the Manchester Literary and Philosophical Society vol. 41, no. 15, 1897; *Mathematical and physical papers*, vol. 5, pp. 256–277.

For $0 < x < na$ it is assumed that

$$\xi_r = Ce^{i(kct - r\theta)} + De^{i(kct + r\theta)}$$

and for $x > na$

$$\xi = Be^{ik(ct - x)}.$$

The kinematical conditions at $x = 0$ and $x = na$ are

$$1 + A = C + D, \qquad Be^{-inka} = Ce^{-in\theta} + De^{in\theta}$$

while the dynamical equations are

$$1 - A = Ci \csc (ka)[\cos (ka) - e^{i\theta}] + iD \csc (ka)[\cos (ka) - e^{-i\theta}],$$

$$Be^{-inka} = iCe^{-in} \csc (ka)[e^{-i\theta} - \cos (ka)]$$
$$+ iDe^{in\theta} \csc (ka)[e^{i\theta} - \cos (ka)]$$

and from these equations it is found that

$$B = \frac{\sin \theta \sin (ka) \exp [(n + 1)ika]}{\sin \theta \sin (ka) \cos (n + 1)\theta + i[1 - \cos \theta \cos (ka)] \sin (n + 1)\theta},$$

$$A = \frac{i(\cos \theta - \cos (ka)) \sin (n + 1)\theta e^{ika}}{\sin \theta \sin (ka) \cos (n + 1)\theta + i[1 - \cos \theta \cos (ka)] \sin (n + 1)\theta}.$$

The intensities of the reflected and transmitted waves are respectively

$$R = [(\cos \theta - \cos (ka))^2 \sin^2 (n + 1)\theta]/H,$$

$$T = [\sin^2 \theta \sin^2 (ka)]/H,$$

where $H = \sin^2 \theta \sin^2 (ka) \cos^2 (n+1)\theta + (1 - \cos \theta \cos (ka))^2 \sin^2 (n+1)\theta$. When n is large slight changes in ka and θ will make $\cos (n+1)\theta$ and $\sin (n+1)\theta$ vary considerably. Mean values of R and T for values of λ close to $2\pi/k$ may be found by integration as in Kirchhoff's *Optik* (p. 165) and in Lamb's paper by using the well known formula

$$\int_0^{\pi/2} \frac{d\phi}{\alpha^2 \cos^2 \phi + \beta^2 \sin^2 \phi} = \pi/2\alpha\beta.$$

Consequently

$$\overline{T} = \left| \frac{\sin \theta \sin (ka)}{1 - \cos \theta \cos (ka)} \right|, \qquad \overline{R} = 1 - \overline{T},$$

$$\overline{T} = \frac{1 - q^2}{1 + q^2}, \qquad \overline{R} = \frac{2q^2}{1 + q^2}$$

where

$$q = \left| \frac{\sin (ka/2 - \theta/2)}{\sin (ka/2 + \theta/2)} \right|.$$

This quantity q is the coefficient of reflection obtained by Lamb in his analysis of the case in which waves travelling towards 0 along the unweighted portion of the string are reflected from the weighted portion on which weights evenly spaced extend to infinity. This case is similar to the problem considered by Godfrey.

An interesting variation of the present problem is one in which the unweighted portions of the string are replaced by weighted portions in which the weights are evenly spaced but at intervals differing in length from the intervals in the intermediate portion of the string. If the intervals in this intermediate portion are greater than in the rest of the string the problem is analogous to that of waves passing from a dense medium into a plate composed of a light substance. This problem is of interest for both longitudinal and transverse vibrations.

4.7. **The reactance theorem and its generalizations.** In 1886 the late Lord Rayleigh[97] gave a dynamical theorem which in modern terminology is a theorem relating to the driving point impedance in a chain like dynamical system. Applications to electrical networks were considered. In 1908 the theorem was extended by the present author[98] to a linear integral equation

$$u(s) = U(s) - \lambda \int_0^1 k(s, t)U(t)dt$$

in which $k(s, t)$ is a real symmetric kernel. If

$$w(\lambda) = \int_0^1 u(s)U(s)ds$$

the theorem then states that the zeros and poles of the function $w(\lambda)$ are all positive when the kernel is such that for any real nonvanishing function $x(s)$ the double integral

$$(xkx) = \int_0^1 \int_0^1 x(s)k(s, t)x(t)dsdt$$

[97] Lord Rayleigh, *The reaction upon the driving-point of a system executing forced harmonic oscillations of various periods, with applications to electricity*, Philosophical Magazine (5) vol. 21 (1886) pp. 369–381; *Scientific papers*, vol. 2, pp. 475–485.

[98] H. Bateman, *The reality of the roots of certain transcendental equations occurring in the theory of integral equations*, Trans. Cambridge Philos. Soc. vol. 20 (1908) pp. 371–382; *Notes on integral equations* III, *the homogeneous integral equation of the first kind*, Messenger of Mathematics vol. 39 (1909) pp. 6–19.

is positive. The derivative $w'(\lambda)$ is then positive for all real values of λ, becoming infinite, if at all, only at certain values of λ at which the homogeneous integral equation (with $u = 0$) can have a nonvanishing solution.

It is the analogue of this theorem for a system of linear algebraic equations

$$u_r = U_r - \lambda \sum_{s=1}^{n} k_{r,s} U_s, \qquad r = 1, 2, \cdots, n,$$

which is essentially the theorem given by Rayleigh. The real constants $k_{r,s}$ are then supposed to be symmetric ($k_{r,s} = k_{s,r}$) and such that the quadratic form

$$(xkx) = \sum_{r,s=1}^{n} x_r k_{r,s} x_s$$

is positive for all nonvanishing sets of real quantities x_r. The function $w(\lambda)$ is then defined by the equation

$$w(\lambda) = \sum_{r=1}^{n} u_r U_r$$

and its derivative is positive for all real values of λ. This indicates that the zeros and poles of the rational function $w(\lambda)$ occur alternately.

In 1922 G. A. Campbell[99] gave an expression for the driving-point impedance Z of a non-dissipative reactance network

$$Z = R + iX = iM \frac{f(f_2^2 - f^2) \cdots (f_{2n}^2 - f^2)^m}{(f_1^2 - f^2) \cdots (f_{2n-1}^2 - f^2)}$$

in which M is a positive constant, $f_2, f_4, \cdots$ are constants representing resonant frequencies, $f_1, f_3, \cdots$ are constants representing anti-resonant frequencies and the exponent m of the last factor in the numerator is 1 or 0 according as a resonant or anti-resonant frequency is the last member when the frequencies f_r are arranged in order of increasing magnitude

$$0 \leqq f_1 < f_2 < f_3 < f_4 \cdots .$$

This type of arrangement is possible because $dX/df > 0$ and so the zeros and poles of X occur alternately. The theorem may be derived from the previous theorem for a system of algebraic equations by

---

[99] G. A. Campbell, *Physical theory of the electric wave-filter*, Bell System Technical Journal vol. 1 (1922) pp. 30–31.

taking all but one of the quantities u_r to be zero. There is a corresponding theorem for the more general set of linear equations

$$u_r = \sum_{s=1}^{n} h_{r,s} U_s - \lambda \sum_{s=1}^{n} k_{r,s} U_s, \quad r = 1, 2, \cdots, n,$$

where the coefficients $h_{r,s}$, $k_{r,s}$ are real symmetric quantities such that the quadratic forms (xhx), (xkx) are positive for nonvanishing sets of quantities x_r. There is also a similar theorem for the integral equation of the first kind

$$u(s) = \int_0^1 h(s, t) U(t) dt - \lambda \int_0^1 k(s, t) U(t) dt$$

in which $h(s, t)$, $k(s, t)$ are real symmetric kernels such that (xhx) and (xkx) are positive for any nonvanishing real continuous function $x(z')$. In all these theorems the function $w(\lambda)$ increases continually with λ but if the conditions imposed on the function k or the coefficients $k_{r,s}$ are relaxed so that k is merely real and symmetric then the function $\lambda w(\lambda)$ increases with λ and there are zeros and poles of this function which occur alternately but are not all positive. The theorem that $w(\lambda)$ or $\lambda w(\lambda)$ increases with λ may be proved in many ways but one way is to use partial fractions. A number of theorems may be proved in this way, some of them are well known in the theory of algebraic equations and some have found useful applications in the theory of electric filters.

The reactance theorem has been proved by many writers. Zobel[100] gives a proof by induction while Foster[101] bases a proof on dynamical theorems of Routh, Rayleigh and Webster. Foster also calls attention to a connection with algebraic equations whose roots are all pseudonegative (that is, with negative real parts). In Routh's work, for instance, such an equation $F(z) = 0$ is expressed in the form $F(z) = E(z) + O(z) = A(z^2) + zB(z^2)$ where $E(z)$ is an even function of z and $O(z)$ is an odd function of z and it is found that the roots of the equations $A(x) = 0$, $B(x) = 0$ are all negative and occur alternately. The converse of this theorem has already been considered in section 2.2.

Foster also considers the driving-point impedance of a general network and connects it with a theorem associated with three positive definite quadratic forms which may be generalized as follows:

[100] O. J. Zobel, *Theory and design of uniform and composite electric wave-filters*, Bell System Technical Journal vol. 2 (1923) pp. 35–36.

[101] R. M. Foster, *A reactance theorem*, ibid. vol. 3 (1924) pp. 259–267; *Theorems regarding the driving-point impedance of two-mesh circuits*, ibid. vol. 3 (1924) pp. 651–685.

Consider the system of linear algebraic equations

$$u_r = \sum_{s=1}^{n} (g_{r,s} + \lambda h_{r,s} + \lambda^2 k_{r,s}) W_s, \quad r = 1, 2, \cdots, n,$$

in which the coefficients $g_{r,s}$, $h_{r,s}$, $k_{r,s}$ are real symmetrical quantities such that $g_{r,s} = g_{s,r}$, $h_{r,s} = h_{s,r}$, $k_{r,s} = k_{s,r}$ and that (xgx), (xhx), (xkx) are positive for any real nonvanishing set of quantities x. Then the zeros and poles of the function $w(\lambda) = \sum_{r=1}^{n} u_r W_r$ are all pseudo-negative.

When λ is a complex quantity $\sigma + i\tau$ and $W_r = U_r + iV_r$, where σ, τ, U_r and V_r are all real, the conjugate complex quantity to W_r, namely $W_r^* = U_r - iV_r$, arises when λ is replaced by $\lambda^* = \sigma - i\tau$ and the zeros of $w(\lambda)$ will be pseudo-negative when the zeros of

$$w^*(\lambda) = \sum_{r=1}^{n} u_r W_r^* = (W^* g W) + \lambda(W^* h W) + \lambda^2(W^* k W)$$

are all pseudo-negative. When, however, the right-hand side is resolved into its real and imaginary parts and each of these equated to zero it is seen at once that σ must be negative and a similar argument is applicable when $u = 0$ and the possible values of λ are, perhaps, poles of $w(\lambda)$. The theorem may be extended to the zeros and poles of the function $w(\lambda) + P$ where P is a positive constant and may also be extended to certain integral equations of type

$$u(s) = \int_{0}^{c} [g(s, t) + \lambda h(s, t) + \lambda^2 k(s, t)] W(t) dt$$

where $g(s, t)$, $h(s, t)$, $k(s, t)$ are real symmetric functions such that (xgx), (xhx), (xkx) are positive for any nonvanishing function $x(s)$ continuous in the range $0 \leqq s \leqq c$.

Proofs and extensions of the reactance theorem have been given by Cauer,[102] Baerwald,[103] Epheser and Glubrecht[104] and many other writ-

[102] W. Cauer, *Die Verwirklichung von Wechselstrom Widerstanden vorgeschrieben Frequenzabhängigkeit*, Dissertation, Technische Hochschule, Berlin, 1926, Arkiv för Elektrot. vol. 17 (1926) pp. 355–388; *Ein Reaktanztheorem*, Sitzungsbericht der Akademie der Wissenschaften, Berlin, 1931, pp. 673–681; *Ein Satz über zwei zusammenhangen Hurwitzsche Polynome*, Sitzungbericht der Berlin Mathematische Gesellschaft vol. 27 (1928) pp. 25–31; *Wechstromschaltingen*, Akademie der Wissenschaften, Leipzig, 1941.

[103] H. G. Baerwald, *Ein einfacher Beweis der Reaktanztheorems*, Elektrische Nachrichten Technik vol. 7 (1930) pp. 331–332.

[104] H. Epheser and H. Glubrecht, *Die Grundlagen der Siebschaltungstheorien*, ibid. vol. 17 (1940) pp. 169–192. See also K. Franz, *Eine Verallgemeinerung des Fosterschen Reaktanztheorems auf beliebige Impedanzen*, Elektrische Nachrichten Technik vol. 20 (1943) pp. 113–115.

ers. There are also extensions to the case in which the quadratic forms (xgx), (xhx), (xkx) contain an infinite number of variables and to the case in which the range of integration 0 to c is infinite in the last mentioned theorem. Extensions to the case of Hermitian forms are also to be considered.

4.8. The sorting of vibrations by means of filters.

It is often important to separate desirable vibrations from the undesirable ones particularly in electric and hydraulic transmission lines, but the problem of separation is important in acoustical and mechanical systems. In acoustics sound waves of a particular frequency or range of frequencies may be needed for experimental work as in the determination of times of reverberation in an auditorium. The elimination of noise is an important requirement in some buildings and laboratories. The absorption of the vibrations produced by engines and heavy machinery provides many mechanical problems.

The filtering properties of a system are generally examined by first finding the behavior of the system in a sinusoidal type of vibration, the so-called steady state, but it is important to know also the somewhat different response of the system to transient disturbances. This second problem is generally much harder than the first. Transient oscillations in electric wave filters were, however, studied by John Carson and Otto Zobel[105] in 1923. The building up of sinusoidal currents in loaded electric lines had been investigated by Carson[106] in 1917 and by Clark[107] and Kupfmüller[108] in 1923. Carson's analysis is

---

[105] J. R. Carson and O. J. Zobel, *Transient oscillations in electric wave filters*, Bell System Technical Journal vol. 2 (1923) pp. 1–52.

[106] J. R. Carson, *Theory and calculation of variable electrical systems*, Physical Review (2) vol. 17 (1921) pp. 116–134; *General expansion theorem for the transient oscillations of a connected system*, ibid. vol. 10 (1917) pp. 217–225; *Theory of the transient oscillations of electrical networks and transmission systems*, Transactions of the American Institute of Electrical Engineers vol. 38 (1919) pp. 345–427, discussion by M. I. Pupin and A. H. Cowles, pp. 462–464. See also T. C. Fry, *The solution of circuit problems*, Physical Review (2) vol. 14 (1919) pp. 115–136; J. R. Carson, *The building up of sinusoidal currents in long periodically loaded lines*, Bell System Technical Journal vol. 3 (1924) pp. 558–566; F. Pollaczek, *Theory of the switching-on process of multimesh artificial lines*, Elektrishe Nachrichten Technik vol. 2 (1925) pp. 197–226.

[107] Alva B. Clark, *Telephone transmission over very long cable circuits*, Transactions of the American Institute of Electrical Engineers vol. 42 (1923) pp. 86–97, discussion by J. J. Pilliod.

[108] K. Küpfmüller, *Free oscillation, echo effect, and influence of temperature in telephony over long Pupinised cables*, Telegraphen- und Fernsprech-Technik vol. 12 (1923) pp. 53–60; *Über Beziehungen zwischen Frequenzcharakteristiken und Ausgleichs vorgängen in linearen Systemen*, Elektrische Nachrichten Technik vol. 5 (1928) pp. 18–32, 459.

based largely on the use of Fourier integrals. For the simpler problem of transients in mechanically loaded lines when use can be made of partial differential equations or of differential difference equations with constant coefficients there is a powerful method of influence functions which has been used effectively by Koppe, Havelock, Schrödinger, Pollaczek and the present writer.[109] Effective use can also be made of the inverse Laplace transformation as suggested by the writer in 1910.[110] In 1921 at the present writer's suggestion,[111] this method was adopted by Carson for work in electrical theory in which the so-called Duhamel integrals and operational methods of Heaviside were being used.

4.9. Transverse vibrations of an infinite light string uniformly loaded at regular intervals. The advantages of the methods mentioned in the last section may be seen by considering the function

$$x_n(t) = \int_0^t J_{2n}[2c(t-s)]f(s)ds$$

in which the integrand of this Poisson-Duhamel integral involves the influence function for the infinite loaded line. When n is a positive or negative integer it is readily seen that $x_n(t)$ satisfies the homogeneous equation

(A) $$x_n''(t) = c^2[x_{n+1}(t) + x_{n-1}(t) - 2x_n(t)]$$

but when $n = 0$ it satisfies the nonhomogeneous equation

$$x_0''(t) = f'(t) + c^2(x_1 + x_{-1} - 2x_0)$$

---

[109] Some references to the literature are given in a paper by the author, *Some simple differential difference equations and the related functions*, Bull. Amer. Math. Soc. vol. 49 (1943) pp. 494–512.

[110] H. Bateman, *The solution of a system of differential equations occurring in the theory of radio-active transformations*, Proc. Cambridge Philos. Soc. vol. 15 (1910) pp. 423–427; *The solution of linear differential equations by means of definite integrals*, Trans. Cambridge Philos. Soc. vol. 21 (1909) pp. 171–196; *Report on the history and present state of the theory of integral equations*, Report of the British Association for the Advancement of Science, 1911, 80 pp.; *An integral equation occurring in a mathematical theory of retail trade*, Messenger of Mathematics vol. 49 (1919–1920) pp. 134–137.

[111] See a letter written by the author to J. R. Carson after the appearance of the latter's paper in Physical Review (1921). The method of the inverse Laplace transformation was recommended also by the author to M. D. Hersey in answer to a letter written during the last war regarding some equations occurring in the theory of recording instruments. The method has been taught by the writer for nearly forty years. Simeon Denis Poisson was probably the first man to use the method for the treatment of differential difference equations in his memoir *Sur la distribution de la Chaleur dans les corps solides*, J. École Polytech. (1) vol. 12 (1923) pp. 1–144, 249–403 (especially pp. 29–34).

and so the motion of the particle numbered 0 is forced while the motion of the other particles is free.

Following the lead of A. E. Heins[112] who has used the method of the inverse Laplace transformation for the classical equation we form (A) the function

$$X_n(z) = \int_0^\infty e^{-zt} x_n(t) dt$$

and a difference equation for $X_n(z)$ can be readily found. In the particular case when $f(t) = \sin(2at)$ the multiplication theorem for integrals of Laplace's type gives the relation

$$X_n(z) = 2a(z^2 + 4a^2)^{-1}(z^2 + 4c^2)^{-1/2}[T(z)]^{|2n|}$$

where

$$2cT(z) = (z^2 + 4c^2)^{1/2} - z.$$

From this expression for $X_n(z)$ the behavior of $x_n(t)$ for large positive values of t may be derived by the method of Mellin and Haar explained in my paper in this Bulletin. Except for a damped initial motion the motion is eventually the steady state of periodic oscillation given by

$$x_n(t) \sim (a^2 - c^2)^{-1/2}[e^{2iat}\{T(2ia)\}^{2n} \pm e^{-2iat}\{T(-2ia)\}^{2n}].$$

When $a > c$ the right-hand side is real when the $+$ sign is used, when $a < c$ the negative sign must be taken to make the right-hand side real. Another interesting example is that in which the pth particle differs in mass from all the others so that

$$x_n'' = c^2(x_{n+1} + x_{n-1} - 2x_n), \qquad n \neq p,$$
$$x_p'' = a^2(x_{p+1} + x_{p-1} - 2x_p), \qquad a \neq c, \ p > 0.$$

A solution for the case in which $x_0(0) = 1$, $x_n(0) = 0$, $n \neq 0$, $x_n'(0) = 0$ is

$$x_n(t) = J_{2n}(2ct) + (1 - c^2/a^2) \int_0^t J_{2n-2p}[2c(t-s)]x_p'(s)ds$$

and

$$X_p(z) = \frac{\{T(z)\}^{|2p|}}{(z^2 + 4c^2)^{1/2} - (1 - c^2/a^2)z}.$$

[112] A. E. Heins, *On the solution of linear difference differential equations*, Journal of Mathematics and Physics, Massachusetts Institute of Technology, vol. 19 (1940) pp. 153–157.

The resulting estimate of $x_p(t)$ for large positive values of t is for $c^2 > 2a^2$

$$x_p(t) \sim \frac{a^2(c^2 - a^2)}{c^2(c^2 - 2a^2)}\, (1 - 2a^2/c^2)^{-|p|} \exp\left[- \frac{2a^2 t}{(c^2 - 2a^2)^{1/2}}\right].$$

When $c^2 > 2a^2$ the motion ultimately decays according to an exponential law. When $c^2 < 2a^2$ the function $X_p(z)$ is infinite for two imaginary values of z and $x_p(t)$ is ultimately periodic. When $c^2 = 2a^2$ the solution is simply

$$x_p(t) = (1/2ct)J_{|2p|+1}(2ct).$$

CALIFORNIA INSTITUTE OF TECHNOLOGY

THE WORK OF LYAPUNOV AND POINCARÉ

by Richard Bellman and Robert Kalaba

IN lieu of presenting a particular paper of either of these two masters to whom modern control theory owes so much, let us attempt to describe briefly some of the guiding ideas of their research in this field. The basic problem to which they addressed themselves was that of determining the qualitative behavior of the solutions of nonlinear differential equations without the luxury of explicit analytic representations. Consider, in particular, a version of the general question of stability of equilibrium. Given a system of differential equations,

$$\frac{dx_i}{dt} = g_i(x_1, x_2, \ldots, x_n), \quad i = 1, 2, \ldots, n, \tag{1}$$

in vector notation,

$$\frac{dx}{dt} = g(x), \tag{2}$$

possessing the null solution $x = 0$, we wish to know under what conditions the solution of

$$\frac{dx}{dt} = g(x), \quad x(0) = c, \tag{3}$$

approaches the null solution as $t \to \infty$. In physical terms, this means that the system described by the equations returns to equilibrium after having been subject to an initial disturbing force, represented by the initial condition $x(0) = c$.

An obvious starting point is the case where the initial disturbances are small and the components of $g(x)$ are analytic in the components of x in the neighborhood of $x = 0$. The equation in (3) can then be written in the form

$$\frac{dx}{dt} = Ax + h(x), \tag{4}$$

where A is a constant matrix, and all components of $h(x)$ are of second order or higher in x. It is tempting then to suppose that the asymptotic behavior of the solutions of the nonlinear differential equation will closely resemble that of the solutions of the linear equation

$$\frac{dx}{dt} = Ax, \tag{5}$$

which can, of course, be solved explicitly. From this explicit solution it follows readily that a necessary and sufficient condition that all solutions of (5) approach zero as $t \to \infty$ is that all of the characteristic roots of A have negative real parts. In this case, we call A a *stability matrix*. The classical result of Lyapunov and Poincaré is:

THEOREM. *If*
(a) $\|g(x)\| = o(\|x\|)$ *as* $\|x\| \to 0$, *where* $\|x\| = \sum_i |x_i|$, $\tag{6}$
(b) *A is a stability matrix,*
then all solutions of (3) for which $\|c\|$ is sufficiently small approach zero as $\to \infty$.

For a variety of proofs under various assumptions, see:

A. Lyapunov, *Annals of Math. Studies*, 1947.

H. Poincaré, *Méthodes nouvelles de la mécanique céleste.* New York: Dover Publications, Inc., 1957.

O. Perron, *Math. Zeit.*, Vol. 29, 1929, pp. 129–160.

E. Cotton, "Approximations successives et les équations différentielles," *Mém. Sci. Fasc.*, Vol. 28. Paris: Hermann et Cie., 1928.

R. Bellman, "On the Boundedness of Solutions of Nonlinear Differential and Difference Equations," *Trans. Amer. Math. Soc.*, Vol. 62, 1947, pp. 357–386.

R. Bellman, *Stability Theory of Differential Equations.* New York: McGraw-Hill Book Company, Inc., 1953.

The assumptions of analytic behavior, constant A, and small initial disturbance are quite restrictive. To obtain stability criteria under more inclusive conditions, different techniques of far-reaching import were introduced, the "second method" of Lyapunov and the geometric method of Poincaré.

The celebrated second method of Lyapunov is based upon the following idea. Let $V(x)$ be a scalar function of $x = x(t)$, a solution of (2). Then

$$\frac{d}{dt} V(x) = \sum_{i=1}^{N} \frac{\partial V}{\partial x_i} \frac{dx_i}{dt} = \sum_{i=1}^{N} \frac{\partial V}{\partial x_i} g_i, \tag{7}$$

upon referring to (1). If one can choose the function $V(x)$ with the property that

$$\sum_{i=1}^{N} \frac{\partial V}{\partial x_i} g_i \leq -kV, \tag{8}$$

where k is a constant, for all x, then $V(x) \leq V(x(0))e^{-kt}$ for $t \geq 0$. If, in addition, $V(x)$ is positive definite, then $V \to 0$ as $t \to \infty$ entails $x \to 0$ as $t \to 0$: stability.

The idea is quite simple, but the selection of $V(x)$ requires a good bit of ingenuity and effort. For an account of these matters, see the original memoir by Lyapunov, cited above, and

J. P. LaSalle and S. Lefschetz, *Stability by Liapunov's Direct Method with Applications*. New York: Academic Press, Inc., 1961.

A. M. Letov, *Stability in Nonlinear Control Systems*. Princeton: Princeton University Press, 1961.

In a recent paper, Letov has shown that Lyapunov functions may be generated by means of control theory and dynamic programming.

Let us note that a great deal of research on stability is going on in the field of mathematical economics, carried on by Arrow, Hurwicz, Uzawa, and others; see for example:

H. Uzawa, "The Stability of Dynamic Processes," *Econometrica*, Vol. 29, 1961, pp. 617–631.

Other references will be found there.

The geometric techniques of Poincaré are based upon the observation that a solution of (1) represents a curve $x = x(t)$ in phase space. Hence, certain intrinsic properties of curves and surfaces can be used to establish properties of solutions. The method has been most successful in dealing with two-dimensional systems, or second order nonlinear differential equations, such as the famed equation of van der Pol (which we shall discuss again below)

$$u'' + \lambda(u^2 - 1)u' + u = 0. \tag{9}$$

The reason for the success of geometric methods in dealing with equations of this type is due to the fact that the uniqueness theorem for solutions of equations of the form

$$u'' + g(u, u') = 0 \tag{10}$$

effectively prevents two solution curves from crossing in the phase plane (u, u'). Alternatively, one could say that the success is due to the Jordan curve theorem.

Furthermore, a closed curve in this plane represents a periodic solution of the original equation

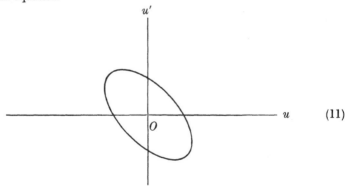

$$\tag{11}$$

It follows that the periodic solutions furnish vital information as to the whereabouts of particular nonperiodic solutions. Thus, for example, a solution starting outside of the oval representing the periodic solution can never approach the origin.

Since a periodic solution can be regarded as equivalent to a fixed-point of a transformation, powerful topological techniques due to Birkhoff-Kellogg, Schauder-Leray, Lefschetz, and others can be used to establish the existence of periodic solutions, and even to estimate their number.

In recent years, the concept of periodic solution has been generalized to that of periodic surface; see:

S. Diliberto, "An Application of Periodic Surfaces (Solution of a Small Divisor Problem)," *Contributions to the Theory of Nonlinear Oscillations*, Annals of Math. Studies No. 36. Princeton: Princeton Univ. Press, Vol. 3, 1956, pp. 257–259.

N. N. Bogoliubov and Yu. A. Mitropolskii, *Asymptotic Methods in the Theory of Nonlinear Oscillations*, Moscow (revised edition 1958). (English editions will be available soon.)

N. N. Bogoliubov and Yu. A. Mitropolskii, "The Method of Integral Manifolds in Nonlinear Mechanics," *Proc. Symp. Nonlinear Osc.*, Kiev, 1961.

J. K. Hale, "Integral Manifolds of Perturbed Differential Systems," *Ann. Math.*, Vol. 73, 1961, pp. 496–531.

But it is not to be expected that results of equal simplicity and power can be established.

The concept of stability has been extended to many other types of functional equations. For differential-difference equations, see:

E. M. Wright, "The Linear Difference-differential Equation with Asymptotically Constant Coefficients," *Amer. J. Math.*, Vol. 70, 1948, pp. 221–238,

as an example of what can be obtained, and for further results and references:

R. Bellman and K. L. Cooke, *Differential-difference Equations.* New York: Academic Press, Inc., 1962.

A. D. Myskis, *Linear Differential Equations with Retarded Argument* (German translation, Berlin: Deutsche Verlag Wiss., 1955).

N. Krasovskii, *Some Problems in the Theory of the Stability of Motion*, Moscow, 1959.

For parabolic partial differential equations, see:

R. Bellman, "On the Existence and Boundedness of Solutions of Nonlinear Partial Differential Equations of Parabolic Types," *Trans. Amer. Math. Soc.*, Vol. 64, 1948, pp. 21–44.

R. Narasimhan, "On the Asymptotic Stability of Solutions of Parabolic Differential Equations," *J. Rat. Mech. and Anal.*, Vol. 3, 1954, pp. 303–319.

A. Friedman, "Asymptotic Behavior of Solutions of Parabolic Equations," *J. Rat. Mech. and Anal.*, Vol. 8, 1959, pp. 387–392.

Other references will be found there.

ON THE CONDITIONS UNDER WHICH AN EQUATION HAS ONLY ROOTS WITH NEGATIVE REAL PARTS*

by A. Hurwitz

As pointed out in our previous discussion, the problem of the behavior over time of a solution of the linear differential equation

$$u^{(n)} + a_1 u^{(n-1)} + \cdots + a_n u = 0, \tag{1}$$

where the coefficients a_i are independent of time, depends upon the location of the roots of the polynomial

$$\lambda^n + a_1 \lambda^{n-1} + \cdots + a_n = 0. \tag{2}$$

In particular, if all solutions of (2) have negative real parts, all solutions of (1) will approach zero as $t \to \infty$. The question of great importance in the design of mechanical and electrical systems is that of determining the location of the roots of (2) by examining the coefficients, without actually solving the equation numerically. Using the classical techniques of Cauchy and Hermite, Hurwitz in a very ingenious fashion derived determinantal criteria which could very easily be applied. Independently, Routh derived equivalent criteria. Since then a great deal of effort has been devoted to questions of this nature because of their fundamental role in circuit synthesis. In general, stability is a necessary and sufficient condition for physical realizability.

Similarly, the asymptotic behavior of the general solution of a differential-difference equation such as

$$u'(t) = a_1 u(t) + a_2 u(t - 1), \tag{3}$$

depends upon the location of the zeros of an exponential polynomial

$$\lambda = a_1 + a_2 e^{-\lambda}. \tag{4}$$

Equations of this type arise in the study of time lags, and we shall encounter them again in our discussion of the pioneering work of Minorsky. Con-

---

* Translated by Howard G. Bergmann from "Ueber die Bedingungen, unter welchen eine Gleichung nur Wurzeln mit negativen reellen Theilen besitzt," *Mathematische Annalen*, Vol. 46, 1895, pp. 273–284.

tinuing in the vein of Cauchy, Pontryagin obtained important results concerning the location of zeroes of exponential polynomials of the form

$$P(\lambda) + Q(\lambda)e^\lambda = 0, \tag{5}$$

for the case where $P(\lambda)$ and $Q(\lambda)$ are ordinary polynomials. See the English translation:

L. Pontryagin, "On the Zeros of Some Transcendental Functions," *Translations Amer. Math. Soc.* (2), Vol. 1, 1955, pp. 95–110,

and Chapter 13 of

R. Bellman and K. L. Cooke, *Differential-difference Equations.* New York: Academic Press, Inc., 1962,

where further results and applications will be found.

For an extensive discussion of Routh-Hurwitz theory and related questions of analysis, see:

F. R. Gantmacher, *Applications of the Theory of Matrices.* New York: Interscience Publishers, 1959, Chapter V.

On the Conditions under which an Equation has only Roots with Negative Real Parts

A. HURWITZ

Translated by Howard G. Bergmann

§ 1

AT THE suggestion of my esteemed colleague, A. Stodola, I considered at some length the question, when does an equation of the nth degree with real coefficients

$$a_0 x^n + a_1 x^{n-1} + \cdots + a_n = 0$$

have only roots whose real parts are negative. Although the solution of this question by the methods of Sturm, Liouville, Cauchy and Hermite presents no particular difficulty, nevertheless I wish to communicate the results I obtained, which may perhaps merit some interest as far as applications are concerned because of their simple form.*

The derivation of the result furnishes simultaneously an opportunity to exhibit the method of Hermite-Jacobi in a form which permits a generalization in various directions.

We may evidently restrict ourselves—as we shall here—to the case where the coefficient a_0 is positive; for otherwise we can multiply the left side of the equation by the factor -1. We now form the determinant

$$\Delta_\lambda = \begin{vmatrix} a_1, a_3, a_5, \ldots, a_{2\lambda-1} \\ a_0, a_2, a_4, \ldots, a_{2\lambda-2} \\ 0, \ a_1, a_3, \ldots, a_{2\lambda-3} \\ \cdot \ \ \cdot \ \ \cdot \ \ \cdot \ \ \cdot \ \ \cdot \\ \cdot \ \ \ \cdot \ \ \ \cdot \ \ \ \cdot \ \ \ a_\lambda \end{vmatrix} \qquad (1)$$

where the indices in each row always increase by two, and always decrease by unity in each column. The term a_κ is set equal to 0, if the index κ is negative or larger than n.

---

* Dr. Stodola made use of my result in his paper "The Regulation of Turbines" (*Schweiz. Bauzeitung*, Vol. 23, Nr. 17, 18). These results were applied at the Davos Spa Turbine Plant with brilliant success. The preceding question has also been raised, as Stodola brought to my attention, in Thomson and Tait's *Natural Philosophy* (1886, Part I, p. 390), and its solution called desirable.

Let us now state the principal theorem:

A necessary and sufficient condition that the equation

$$a_0 x^n + a_1 x^{n-1} + \cdots + a_n = 0, \tag{2}$$

with real coefficients in which the coefficient a_0 is assumed to be positive, have only roots with negative real parts, is that the values of the determinants

$$\Delta_1 = a_1, \Delta_2, \Delta_3, \ldots, \Delta_n \tag{3}$$

all be positive.

In connection with this theorem, the following is further to be remarked. Expanding with respect to the elements of the last column, the determinant Δ_n is readily seen to equal $a_n \Delta_{n-1}$.

Therefore the requirement that Δ_{n-1} and Δ_n be positive is equivalent to the requirement that Δ_{n-1} and a_n be positive. The theorem above therefore remains valid if Δ_n is replaced by a_n. Another remark is the following:

Consider the sequence of determinants

$$\Delta_1, \Delta_2, \Delta_3, \Delta_4, \ldots. \tag{4}$$

The terms of this sequence from the $(n+1)$st on vanish identically, for arbitrary values of $a_0, a_1, \ldots, a_n$, since the elements of the last column of Δ_λ are all zero for $\lambda > n$. The conditions of the theorem can therefore be formulated in the following way: The non-identically vanishing terms of the sequence (4) must all be positive. The terms of this sequence can be written out fully,

$$a_1, \begin{vmatrix} a_1, a_3 \\ a_0, a_2 \end{vmatrix}, \begin{vmatrix} a_1, a_3, a_5 \\ a_0, a_2, a_4 \\ 0, a_1, a_3 \end{vmatrix}, \begin{vmatrix} a_1, a_3, a_5, a_7 \\ a_0, a_2, a_4, a_6 \\ 0, a_1, a_3, a_5 \\ 0, a_0, a_2, a_4 \end{vmatrix}, \ldots,$$

and we can therefore immediately form the conditions for each particular value of n.

For example, for the equation of the 4th degree ($n = 4$), the conditions are

$$a_1 > 0, \begin{vmatrix} a_1, a_3 \\ a_0, a_2 \end{vmatrix} > 0, \begin{vmatrix} a_1, a_3, 0 \\ a_0, a_2, a_4 \\ 0, a_1, a_3 \end{vmatrix} > 0, a_4 > 0.$$

Stodola has remarked that a *necessary* condition that equation (2) have only roots with negative real parts is that all the coefficients $a_0, a_1, \ldots, a_n$ be positive. Indeed, if the real parts of all the roots of equation (2) are negative, then every real linear factor of the left side of the equation has the form $x + p$, and every real quadratic factor the form

$$(x + p_1)^2 + p_2^2 = x^2 + p'x + p'',$$

where p, p_1, p_2, p', p'' represent positive quantities. Since the product of polynomials with positive coefficients likewise has positive coefficients, it follows that the left side of equation (2) will have only positive coefficients.

§ 2

Let the entire rational function $f(x)$, whose coefficients may be complex, be subject to the condition that it vanish for no purely imaginary values of x. Let us then designate by N and P the number of zeros of $f(x)$ which have negative and positive real parts, respectively, so that

$$N + P = n, \tag{5}$$

where n stands for the degree of $f(x)$. Now let c be an arbitrary (complex) constant and let

$$cf(x) = \rho \cdot e^{in\phi}, \tag{6}$$

so that ρ represents the absolute value and $\pi\phi$ the arguments of $cf(x)$. The angle ϕ varies continuously with the value of x and in particular decreases

$$N - P = \Delta \tag{7}$$

units if x ranges over purely imaginary values from $+i\infty$ to $-i\infty$. This is seen immediately, by making use of the customary geometrical representation of complex numbers. We follow the variation of the argument of the individual linear factor of $f(x)$. From (5) and (7), we now have

$$N = \frac{n + \Delta}{2}, \qquad P = \frac{n - \Delta}{2}. \tag{8}$$

The determination of Δ is now reduced in well-known fashion to that of the evaluation of a Cauchy index.* In general, we understand by the index of a function R, which has a definite real value at every point of a line L traversed in a specific sense, a number to be constructed in the following manner. We assign to each point of L, at which R is infinite, the number 0 or $+1$ or -1, according as R does not change sign as the point is passed by, or changes from negative to positive values, or from positive to negative. The index of R with respect to the line L is then the sum of all the numbers assigned at the points where R is infinite. We are tacitly assuming that R becomes infinite and changes sign only at a finite number of points, and that $1/R$ is continuous in the neighborhood of such points.

Now let z be a real variable and write

$$cf(-iz) = U + iV, \tag{9}$$

where U and V represent entire functions of z with real coefficients. If we now set

$$\frac{V}{U} = R(z), \tag{10}$$

---

* *Journal de l'école polytechnique*, XV (1837). The Cauchy index is included as a special case of the notion of the characteristic of a system of functions, introduced by Kronecker (*Monatsberichte der kgl. preussischen Akademie der Wissenschaften*, 1869).

we then have

$$\phi = \frac{1}{\pi} \text{ arc tan } R(z).$$

From this equation it follows that Δ is the same as the index of $R(z)$ with respect to the real z-axis, traversed in the direction of increasing z, where the real z-axis is regarded as a line closed at infinity. In what follows we assume that $R(z)$ does not become infinite for $z = \infty$, which is clearly permissible, since we can choose the constant c arbitrarily.

§ 3

Now let $R(z)$ be any rational function of z with real coefficients which remains finite for $z = \infty$.

The index of $R(z)$ (with respect to the axis of reals, traversed in the direction of increasing z) can be determined, as is well known, by the Sturm division algorithm, or by the Hermitian reduction to a quadratic form, whose signature is equal to the desired index. By "signature" of a quadratic form with real coefficients we mean, following Frobenius,* the difference between the number of positive and of negative squares which occur in the representation of the form as a sum of the fewest possible squares of real linear functions.

We are led to this method of Hermite's for determining the index of $R(z)$ in the following way. Let

$$\Theta(z) = y_0 + y_1 z + y_2 z^2 + \cdots + y_{m-1} z^{m-1} \tag{11}$$

designate an entire rational function of z, whose coefficients can be regarded as arbitrary parameters; then the integral

$$F_m = \frac{1}{2\pi i} \int R(z)[\Theta(z)]^2 \, dz, \tag{12}$$

taken around a curve enclosing all the poles of $R(z)$, represents a quadratic form of the parameters $y_0, y_1, \ldots, y_{m-1}$, which can be obtained easily as the coefficient of $1/z$ in the expansion of $R(z)[\Theta(z)]^2$ in ascending powers of $1/z$.† On the other hand, the integral is equal to the sum of the residues of

---

* "On the Law of Inertia of quadratic forms" (*Sitzungsberichte der kgl. preussischen Akademie der Wissenschaften*, 1894).

† Instead of the integral (12), we can also consider the integral

$$\frac{1}{2\pi i} \int R(z) \cdot \frac{[\Theta(z)]^2}{(z - \alpha)^{2m}} \, dz$$

taken around the value $z = \alpha$, where α stands for a real value, for which $R(z)$ remains finite. $z = \alpha$ plays the same role for this integral as $z = \infty$ does for the integral (12). In this connection, we have the immediate fact that the index of $R\left(\dfrac{az + b}{cz + d}\right)$ is equal to the index of $R(z)$, supposing that a, b, c, d are real constants whose determinant $ad - bc$ is positive.

$R(z)[\Theta(z)]^2$, which correspond to the poles of $R(z)$. Let $z = a$ be a simple pole of $R(z)$ and let

$$R(a + t) = \frac{c}{t} + c_1 + c_2 + \cdots; \qquad (13)$$

then the residue at $z = a$ is

$$c \cdot [\Theta(a)]^2.$$

If a is real, then the pole $z = a$ contributes $+1$ or -1 to the index of $R(z)$ according as c is positive or negative. If a is imaginary and $\bar{a}$ designates the pole conjugate to a, then the sum of the residues at a and $\bar{a}$ is

$$c[\Theta(a)]^2 + \bar{c}[\Theta(\bar{a})]^2 = (P + iQ)^2 + (P - iQ)^2$$
$$= 2P^2 - 2Q^2,$$

where P and Q are real linear functions. From this follows (for the time being under the assumption that $R(z)$ has only simple poles) the theorem:

If n designates the number of poles of $R(z)$, then the quadratic form F_m can be represented as a sum of n squares, in which the difference between the number of positive squares and the number of negative squares is equal to the index of $R(z)$.

However, this theorem is also valid for the case in which $R(z)$ has poles of any multiplicity, where n is to be understood to be the number of poles, each counted according to its multiplicity.* In order to prove this, let $z = a$ be a λ-fold pole of $R(z)$, and let

$$R(a + t) = \frac{c}{t^\lambda} + \frac{c_1}{t^{\lambda-1}} + \cdots + \frac{c_{\lambda-1}}{t} + \cdots,$$

$$\Theta(a + t) = \Theta_0(a) + \Theta_1(a)t + \Theta_2(a)t^2 + \cdots,$$

where $\Theta_0(a)$, $\Theta_1(a)$, $\ldots$, designate linear forms in the parameters $y_0, y_1, \ldots,$ y_{m-1}. The residue at $z = a$ is then

$$c_{\lambda-1}\Theta_0^2 + 2c_{\lambda-2}\Theta_0\Theta_1 + \cdots + c(2\Theta_0\Theta_{\lambda-1} + 2\Theta_1\Theta_{\lambda-2} + \cdots).$$

According as λ is even or odd, this residue can be put in either the form $\Theta_0\psi_0 + \Theta_1\psi_1 + \cdots + \Theta_{\mu-1}\psi_{\mu-1}$ $(\lambda = 2\mu)$, or the form

$$\Theta_0\psi_0 + \Theta_1\psi_1 + \cdots + \Theta_{\mu-1}\psi_{\mu-1} + c\Theta_\mu^2 \quad (\lambda = 2\mu + 1),$$

where $\psi_0, \psi_1, \ldots$ stand for linear functions of the parameters.

If a is real, then the coefficients of $\Theta_0, \Theta_1, \ldots, \psi_0, \psi_1, \ldots$ are also real, and the residue can be brought into the form

$$[\tfrac{1}{2}(\Theta_0 + \psi_0)]^2 - [\tfrac{1}{2}(\Theta_0 - \psi_0)]^2 + \cdots + [\tfrac{1}{2}(\Theta_{\mu-1} + \psi_{\mu-1})]^2$$
$$- [\tfrac{1}{2}(\Theta_{\mu-1} - \psi_{\mu-1})]^2 \quad (\lambda = 2\mu)$$

---

* That the conclusions with appropriate modifications with reference to the Sturm series also still remain valid when the entire functions under consideration have multiple linear factors, was remarked by Kronecker in his paper "Toward the Theory of the Elimination of a Variable from Two Algebraic Equations" (*Monatsberichte der kgl. preussischen Akademie der Wissenschaften*, 1881).

or the form

$$[\tfrac{1}{2}(\Theta_0 + \psi_0)]^2 - [\tfrac{1}{2}(\Theta_0 - \psi_0)]^2 + \cdots + [\tfrac{1}{2}(\Theta_{\mu-1} + \psi_{\mu-1})]^2$$
$$- [\tfrac{1}{2}(\Theta_{\mu-1} - \psi_{\mu-1})]^2 + c\Theta_\mu^2 \qquad (\lambda = 2\mu + 1),$$

in which it appears as a sum of λ squares of real linear forms.

If λ is even, there occur as many positive squares as negative ones; on the other hand, if λ is odd, there is one more positive or one more negative square, according as c is positive or negative. The discussion of the case in which $z = a$ is complex, is carried through in a similar fashion, and we thus recognize the general validity of the foregoing theorem.

§ 4

If $m > n$, then the quadratic form F_m has a vanishing determinant, since the form can be represented as a sum of n squares, hence as a form in less than m linear combinations of the parameters $y_0, y_1, \ldots, y_{m-1}$. On the other hand, the determinant of the form F_n is different from zero. We can prove this either by showing the identity of this determinant with the resultant of the numerator and denominator of the rational function $R(z)$ written in reduced form (see Sec. 6 below), or else in the following way:

If the determinant of F_n vanished, then we could find values of $y_0, y_1, \ldots, y_{n-1}$, not all vanishing, for which $\partial F_n/\partial y_0, \partial F_n/\partial y_1, \ldots, \partial F_n/\partial y_{n-1}$, that is, for which the integrals

$$\frac{1}{2\pi i} \int R(z) \cdot \Theta(z) \cdot z^\lambda \, dz \qquad (\lambda = 0, 1, \ldots, n-1) \qquad (14)$$

are all zero. If now

$$R(z) \cdot \Theta(z) = G(z) + R_1(z), \qquad (15)$$

where $G(z)$ represents an entire rational function of z, and

$$R_1(z) = R(z) \cdot \Theta(z) - G(z) = \frac{k'}{z} + \frac{k''}{z^2} + \cdots \qquad (16)$$

represents a rational function vanishing for $z = \infty$, then for the vanishing of those integrals it is necessary that

$$k' = k'' = \cdots = k^{(n)} = 0.$$

Therefore, it must be true that $R_1(z)$ vanishes at least to the $(n + 1)$st order at $z = \infty$. However, since $R_1(z)$ can become infinite only at the poles of $R(z)$ (whence at most n times), $R_1(z)$ must vanish identically. The equation resulting from this, $R(z) \cdot \Theta(z) = G(z)$, is, however, impossible, since $\Theta(z)$ is at most of the $(n-1)$st degree, and $R(z)$ has n poles.

§ 5

From the foregoing considerations, we derive the following method for the determination of the index of $R(z)$: Let

$$R(z) = c + \frac{c_0}{z} + \frac{c_1}{z^2} + \frac{c_2}{z^3} + \cdots \tag{17}$$

be the development of $R(z)$ in the neighborhood of $z = \infty$. The coefficient of $1/z$ in the expansion of the product of $R(z)$ and

$$[\Theta(z)]^2 = \sum_{i,\,k} y_i y_k z^{i+k} \qquad (i, k = 0, 1, \ldots, m-1) \tag{18}$$

is therefore

$$F_m = \sum_{i,\,k} c_{i+k} y_i y_k \qquad (i, k = 0, 1, \ldots, m-1), \tag{19}$$

and the determinant of the form F_m can be represented in the form

$$D_m = \begin{vmatrix} c_0, & c_1, & \ldots, & c_{m-1} \\ c_1, & c_2, & \ldots, & c_m \\ \cdot & \cdot & \cdot & \cdot \\ \cdot & \cdot & \cdot & \cdot \\ \cdot & \cdot & \cdot & \cdot \\ c_{m-1}, & c_m, & \ldots, & c_{2m-2} \end{vmatrix}. \tag{20}$$

In the sequence of determinants

$$D_1, D_2, D_3, \ldots \tag{21}$$

all the terms from a certain one on, say D_{n+1}, are now equal to zero, while D_n is different from zero. Therefore n furnishes the number of poles of $R(z)$, and the index of $R(z)$ is equal to the signature of the form F_n.

The signature of the form F_n can in every case be read off from the sign of the non-vanishing determinants among $D_1, D_2, \ldots, D_{n-1}$.[*] In the case where none of these determinants vanishes, F_n can be represented in the form

$$F_n = D_1 u_0^2 + \frac{D_2}{D_1} u_1^2 + \cdots + \frac{D_n}{D_{n-1}} u_{n-1}^2,$$

where u_i is a real linear form in $y_i, y_{i+1}, \ldots, y_{n-1}$. Therefore, the index of $R(z)$ is equal to the difference between the number of positive and the number of negative terms of the sequence

$$D_1, \frac{D_2}{D_1}, \frac{D_3}{D_2}, \ldots, \frac{D_n}{D_{n-1}}.$$

---

[*] Frobenius, *loc. cit.*, p. 410.

This case occurs when the index of $R(z)$ attains its maximum value n. For then F_n is a positive definite form and so also are $F_{n-1}, F_{n-2}, \ldots, F_1$, since the latter forms are obtained from F_n by setting some of the parameters $y_0, y_1, \ldots, y_{n-1}$ equal to zero. Since the determinant of a positive definite form is always positive, we have the theorem:

The index of $R(z)$ has its maximum value n if the determinants $D_1, D_2, \ldots, D_n$ are positive, and only in this case.

§ 6

Now let $R(z)$ be given in the form

$$R(z) = \frac{b_0 z^\nu + b_1 z^{\nu-1} + \cdots + b_\nu}{a_0 z^\nu + a_1 z^{\nu-1} + \cdots + a_\nu}, \tag{22}$$

where the coefficient a_0 is assumed different from zero. The degree ν of the denominator of $R(z)$ is greater than or equal to n, according as the numerator and denominator of $R(z)$ have or do not have a common divisor. We can now transform the determinant D_m (20) into a determinant in which the coefficients $a_0, \ldots, a_\nu,$ $b_0, \ldots, b_\nu$ constitute the elements. This transformation can be worked out with the aid of the following theorem, which can easily be derived from the theorem on multiplication of determinants.

Let

$$\mathfrak{P}_1, \mathfrak{P}_2, \ldots, \mathfrak{P}_m, \ldots \tag{23}$$

be ordinary power series in z, which may be changed through multiplication by

$$\mathfrak{P} = k + k_1 z + k_2 z^2 + \cdots \tag{24}$$

into the new power series

$$\mathfrak{P}'_1, \mathfrak{P}'_2, \ldots, \mathfrak{P}'_m, \ldots, \tag{25}$$

so that $\mathfrak{P}'_m = \mathfrak{P}\mathfrak{P}_m$. From each of the series $\mathfrak{P}_1, \mathfrak{P}_2, \ldots, \mathfrak{P}_m$ (or $\mathfrak{P}'_1, \mathfrak{P}'_2, \ldots, \mathfrak{P}'_m$), eliminate the first m terms and designate by Δ_m (or Δ'_m) the determinant of the m entire functions of z of the $(m-1)$st degree thus generated; then it is easy to see that

$$\Delta'_m = k^m \Delta_m. \tag{26}$$

Now apply this theorem to the following case. Let

$$\frac{b_0 + b_1 z + b_2 z^2 + \cdots}{a_0 + a_1 z + a_2 z^2 + \cdots} = c + c_0 z + c_1 z^2 + \cdots,$$

and the series (23) may be taken as follows:

$$\mathfrak{P}_1 = 1, \mathfrak{P}_2 = c + c_0 z + c_1 z^2 + \cdots, \mathfrak{P}_{2\lambda+1} = z^\lambda \mathfrak{P}_1, \mathfrak{P}_{2\lambda+2} = z^\lambda \mathfrak{P}_2$$
$$(\lambda = 1, 2, \ldots),$$

while the series (24) will be identified with

$$\mathfrak{P} = a_0 + a_1 z + a_2 z^2 + \cdots.$$

The series (25) then are

$$\mathfrak{P}'_1 = \mathfrak{P}, \; \mathfrak{P}'_2 = b_0 + b_1 z + b_2 z^2 + \cdots,$$

$$\mathfrak{P}'_{2\lambda+1} = z^\lambda \mathfrak{P}'_1 \mathfrak{P}'_{2\lambda+2} = z^\lambda \mathfrak{P}'_2 \quad (\lambda = 1, 2, \ldots).$$

If we further substitute in equation (26) the index $2m$ for m, this equation will now give the desired transformation of the determinant D_m; namely

$$a_0^{2m} \cdot D_m = R_m, \tag{27}$$

where R_m denotes the determinant

$$R_m = \begin{vmatrix} a_0, a_1, \ldots, a_{2m-1} \\ b_0, b_1, \ldots, b_{2m-1} \\ 0, \; a_0, \ldots, a_{2m-2} \\ 0, \; b_0, \ldots, b_{2m-2} \\ \cdot \;\; \cdot \;\; \cdot \;\; \cdot \;\; \cdot \\ \cdot \;\; \cdot \;\; \cdot \;\; \cdot \;\; \cdot \\ 0, \; 0, \; \ldots, a_m \\ 0, \; 0, \; \ldots, b_m \end{vmatrix}. \tag{28}$$

This clearly vanishes, as soon as $m > \nu$, since the elements of the last column will all be zero. Accordingly, we use the following procedure for the determination of the index (and at the same time the number n of the poles) of the rational function (22): Construct the sequence of determinants

$$R_1, R_2, \ldots, R_\nu.$$

If R_n is the last non-vanishing term in this sequence, then n gives the number of the poles, or, what is the same thing, the degree of the denominator of $R(z)$, if $R(z)$ has been expressed in reduced form. The index of $R(z)$ will in that case be obtained immediately from the signs of the non-vanishing terms of the sequence $R_1, R_2, \ldots, R_n$.

§ 7

The theorem given in Sec. 1 now follows easily. Let

$$f(x) \equiv a_0 x^n + a_1 x^{n-1} + \cdots + a_n = 0 \tag{29}$$

be an equation with real coefficients. Then

$$i^n f(-iz) = (a_0 z^n - a_2 z^{n-2} + \cdots) + i(a_1 z^{n-1} - a_3 z^{n-3} + \cdots), \tag{30}$$

and the number designated in Sec. 2 by Δ is the index of

$$R(z) = \frac{a_1 z^{n-1} - a_3 z^{n-3} + \cdots}{a_0 z^n - a_2 z^{n-1} + \cdots}. \tag{31}$$

Equation (29) has, as follows from (8) in Sec. 2, all its roots with negative real parts, if and only if $\Delta = n$. It follows that the numerator and denominator of $R(z)$ must be relatively prime. Otherwise, $R(z)$ could be represented as a quotient, with denominator of degree $n' < n$, and the index of $R(z)$ would be at most equal to n'.

The necessary and sufficient condition for equation (29) to have only roots with negative real parts is therefore that the quadratic form

$$F_n = \frac{1}{2\pi i} \int R(z)[\Theta(z)]^2 \, dz \tag{32}$$

be a positive definite form in $y_0, y_1, \ldots, y_{n-1}$. Since $R(z)$ is an odd function of z, F_n can be broken up into two forms, one of which contains only the parameters $y_0, y_2, y_4, \ldots$, the other only the parameters $y_1, y_3, y_5, \ldots$. Indeed, let

$$H(z) = \frac{a_1 z^{\lambda-1} - a_3 z^{\lambda-2} + \cdots}{a_0 z^\lambda - a_2 z^{\lambda-1} + \cdots}, \tag{33}$$

($\lambda = \frac{1}{2}n$ or $(n+1)/2$ according as n is even or odd),

where, evidently,

$$R(z) = z \cdot H(z^2).$$

Furthermore, let us collect in $\Theta(z)$ the even and odd terms in z and then set

$$\Theta(z) = \Theta_0(z^2) + z\Theta_1(z^2).$$

In the integral, introduce (32) $z^2 = \zeta$ as the new variable of integration, and then write z again in place of ζ. We will then obtain the decomposition

$$F_n = \frac{1}{2\pi i} \int H(z)[\Theta_0(z)]^2 \, dz + \frac{1}{2\pi i} \int zH(z)[\Theta_1(z)]^2 \, dz. \tag{34}$$

From this equation we see that the index of $R(z)$ is equal to the sum of the indices of $H(z)$ and $zH(z)$, a fact which can be immediately derived from the definition of index. If we now set up the conditions of Secs. 5 and 6 that F_n, or what amounts to the same thing, that each of the two integrals (34), represents a positive definite form, we will be led, after an easy transformation of the determinants constructed in this manner, to the theorem of Sec. 1.

§ 8

By means of equation (8) of Sec. 2 and the method developed in Sec. 6 for the determination of the index of a rational function, the problem of determining the number of those roots of an equation $f(x) = 0$ which have a negative real part, is solved in all generality under the assumption that the equation has no purely imaginary roots. (This last restriction can, moreover, be dropped if we stipulate that each pure, imaginary root will be counted with multiplicity $1/2$.)   This problem is not essentially different, as the substitution of $-ix$ in place of x shows, from the other, that of determining the number of roots of an equation of nth degree

$$f_1(x) + if_2(x) = 0, \qquad (35)$$

which have a positive imaginary part, where $f_1(x)$ and $f_2(x)$ designate entire functions with real coefficients.   This number will also be given by the first formula (8), and hence by $(n + \Delta)/2$, where by Δ we mean the index of $f_2(x)/f_1(x)$.

Hermite treated the last problem in two papers.*   In conclusion, let us make the further remark: From the notion of index it immediately follows that a rational function $f_2(x)/f_1(x)$ has the index $\pm n$, if and only if the denominator $f_1(x)$ vanishes at n points of the real axis (where $x = \infty$ is considered as a zero if $f(x)$ in case $f_1(x)$ is only of the $(n - 1)$st degree) and if at the same time $f_2(x)$ assumes at every consecutive pair of these points values of opposite sign.   Hence it further follows that the maximum value $\pm n$ of the index of $f_2(x)/f_1(x)$ occurs if and only if each of the equations $f_1(x) = 0$, $f_2(x) = 0$ has n distinct real roots and also the roots of one equation are separated by those of the other.   In particular, therefore, the n roots of equation (35) have all positive-imaginary or all negative-imaginary parts, if and only if the roots of the equations $f_1(x) = 0$, $f_2(x) = 0$ have the properties just mentioned.†

Zürich, December 12, 1894

---

* *Crelle's Journal*, Vol. 52, p. 39, *Bulletin de la société mathématique de France*, Vol. 7, p. 128.

† Biehler, *Crelle's Journal*, Vol. 87, p. 350; Laguerre, *ibid.*, Vol. 89, p. 339.

REGENERATION THEORY*

by H. Nyquist

SINCE the location of the roots and poles of meromorphic functions is of prime importance in the study of stability of linear systems, one finds many results in complex variable theory concerning the numbers of roots and poles in a given region playing a significant role. In particular, if N is the number of zeros and P the number of poles of the function $f(z)$, we have the fundamental formula of Cauchy:

$$N - P = \frac{1}{2\pi i} \int \frac{f'(z)}{f(z)} \, dz,$$

where the integration is over the boundary of the region. A discussion of this theorem, Rouché's theorem, and others can be found in books on analytic function theory, e.g.:

E. C. Titchmarsh, *The Theory of Functions*. London: Oxford University Press, 1939.

The purpose of the following paper is to give a careful discussion of the notion of stability of linear feedback, or "regeneration" systems, together with consideration of steady-state and transient behavior. The main result is the Nyquist criterion for stability, a standard tool of control engineers and the focal point of many subsequent investigations.

---

* From *Bell System Technical Journal*, Vol. 11, 1932, pp. 126–147, by permission of the American Telephone and Telegraph Company.

Regeneration Theory

By H. NYQUIST

Regeneration or feed-back is of considerable importance in many applications of vacuum tubes. The most obvious example is that of vacuum tube
oscillators, where the feed-back is carried beyond the singing point. Another
application is the 21-circuit test of balance, in which the current due to the
unbalance between two impedances is fed back, the gain being increased
until singing occurs. Still other applications are cases where portions of
the output current of amplifiers are fed back to the input either unin-
tentionally or by design. For the purpose of investigating the stability of
such devices they may be looked on as amplifiers whose output is connected
to the input through a transducer. This paper deals with the theory of
stability of such systems.

PRELIMINARY DISCUSSION

WHEN the output of an amplifier is connected to the input through a transducer the resulting combination may be either stable or unstable. The circuit will be said to be stable when an impressed small disturbance, which itself dies out, results in a response which dies out. It will be said to be unstable when such a disturbance results in a response which goes on indefinitely, either staying at a relatively small value or increasing until it is limited by the non-linearity of the amplifier. When thus limited, the disturbance does not grow further. The net gain of the round trip circuit is then zero. Otherwise stated, the more the response increases the more does the non-linearity decrease the gain until at the point of operation the gain of the amplifier is just equal to the loss in the feed-back admittance. An oscillator under these conditions would ordinarily be called stable but it will simplify the present paper to use the definitions above and call it unstable. Now, this fact as to equality of gain and loss appears to be an accident connected with the non-linearity of the circuit and far from throwing light on the conditions for stability actually diverts attention from the essential facts. In the present discussion this difficulty will be avoided by the use of a strictly linear amplifier, which implies an amplifier of unlimited power carrying capacity. The attention will then be centered on whether an initial impulse dies out or results in a runaway condition. If a runaway condition takes place in such an amplifier, it follows that a non-linear amplifier having the same gain for small current and decreasing gain with increasing current will be unstable as well.

126

REGENERATION THEORY   127

STEADY-STATE THEORIES AND EXPERIENCE

First, a discussion will be made of certain steady-state theories; and reasons why they are unsatisfactory will be pointed out. The most obvious method may be referred to as the series treatment. Let the complex quantity $AJ(i\omega)$ represent the ratio by which the amplifier and feed-back circuit modify the current in one round trip, that is, let the magnitude of AJ represent the ratio numerically and let the angle of AJ represent the phase shift. It will be convenient to refer to AJ as an admittance, although it does not have the dimensions of the quantity usually so called. Let the current

$$I_0 = \cos \omega t = \text{real part of } e^{i\omega t} \qquad (a)$$

be impressed on the circuit. The first round trip is then represented by

$$I_1 = \text{real part of } AJe^{i\omega t} \qquad (b)$$

and the nth by

$$I_m = \text{real part of } A^n J^n e^{i\omega t}. \qquad (c)$$

The total current of the original impressed current and the first n round trips is

$$I_n = \text{real part of } (1 + AJ + A^2J^2 + \cdots A^nJ^n)e^{i\omega t}. \qquad (d)$$

If the expression in parentheses converges as n increases indefinitely, the conclusion is that the total current equals the limit of (d) as n increases indefinitely. Now

$$1 + AJ + \cdots A^nJ^n = \frac{1 - A^{n+1}J^{n+1}}{1 - AJ}. \qquad (e)$$

If $|AJ| < 1$ this converges to $1/(1 - AJ)$ which leads to an answer which accords with experiment. When $|AJ| > 1$ an examination of the numerator in (e) shows that the expression does not converge but can be made as great as desired by taking n sufficiently large. The most obvious conclusion is that when $|AJ| > 1$ for some frequency there is a runaway condition. This disagrees with experiment, for instance, in the case where AJ is a negative quantity numerically greater than one. The next suggestion is to assume that somehow the expression $1/(1 - AJ)$ may be used instead of the limit of (e). This, however, in addition to being arbitrary, disagrees with experimental results in the case where AJ is positive and greater than 1, where the expression $1/(1 - AJ)$ leads to a finite current but where experiment indicates an unstable condition.

The fundamental difficulty with this method can be made apparent by considering the nature of the current expressed by (*a*) above. Does the expression cos ωt indicate a current which has been going on for all time or was the current zero up to a certain time and cos ωt thereafter? In the former case we introduce infinities into our expressions and make the equations invalid; in the latter case there will be transients or building-up processes whose importance may increase as *n* increases but which are tacitly neglected in equations (*b*) — (*e*). Briefly then, the difficulty with this method is that it neglects the building-up processes.

Another method is as follows: Let the voltage (or current) at any point be made up of two components

$$V = V_1 + V_2, \tag{f}$$

where *V* is the total voltage, V_1 is the part due directly to the impressed voltage, that is to say, without the feed-back, and V_2 is the component due to feed-back alone. We have

$$V_2 = AJV. \tag{g}$$

Eliminating V_2 between (*f*) and (*g*)

$$V = V_1/(1 - AJ). \tag{h}$$

This result agrees with experiment when $|AJ| < 1$ but does not generally agree when AJ is positive and greater than unity. The difficulty with this method is that it does not investigate whether or not a steady state exists. It simply assumes tacitly that a steady state exists and if so it gives the correct value. When a steady state does not exist this method yields no information, nor does it give any information as to whether or not a steady state exists, which is the important point.

The experimental facts do not appear to have been formulated precisely but appear to be well known to those working with these circuits. They may be stated loosely as follows: There is an unstable condition whenever there is at least one frequency for which AJ is positive and greater than unity. On the other hand, when AJ is negative it may be very much greater than unity and the condition is nevertheless stable. There are instances of $|AJ|$ being about 100 without the conditions being unstable. This, as will appear, accords closely with the rule deduced below.

REGENERATION THEORY 129

NOTATION AND RESTRICTIONS

The following notation will be used in connection with integrals:

$$\int_I \phi(z)dz = \lim_{M \to \infty} \int_{-iM}^{+iM} \phi(z)dz, \tag{1}$$

the path of integration being along the imaginary axis (see equation 9), i.e., the straight line joining $- iM$ and $+ iM$;

$$\int_{s+} \phi(z)dz = \lim_{M \to \infty} \int_{-iM}^{iM} \phi(z)dz, \tag{2}$$

the path of integration being along a semicircle [1] having the origin for center and passing through the points $- iM$, M, iM;

$$\int_C \phi(z)dz = \lim_{M \to \infty} \int_{-iM}^{-iM} \phi(z)dz, \tag{3}$$

the path of integration being first along the semicircle referred to and then along a straight line from iM to $- iM$. Referring to Fig. 1 it

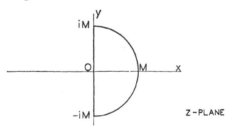

Fig. 1—Paths of integration in the z-plane.

will be seen that

$$\int_{s+} - \int_I = \int_C. \tag{4}$$

The total feed-back circuit is made up of an amplifier in tandem with a network. The amplifier is characterized by the amplifying ratio A which is independent of frequency. The network is characterized by the ratio $J(i\omega)$ which is a function of frequency but does not depend on the gain. The total effect of the amplifier and the network is to multiply the wave by the ratio $A J(i\omega)$. An alternative way of characterizing the amplifier and network is to say that the amplifier is

[1] For physical interpretation of paths of integration for which $x > 0$ reference is made to a paper by J. R. Carson, "Notes on the Heaviside Operational Calculus," *B. S. T. J.*, Jan. 1930. For purposes of the present discussion the semicircle is preferable to the path there discussed.

characterized by the amplifying factor A which is independent of time, and the network by the real function $G(t)$ which is the response caused by a unit impulse applied at time $t = 0$. The combined effect of the amplifier and network is to convert a unit impulse to the function $AG(t)$. Both these characterizations will be used.

The restrictions which are imposed on the functions in order that the subsequent reasoning may be valid will now be stated. There is no restriction on A other than that it should be real and independent of time and frequency. In stating the restrictions on the network it is convenient to begin with the expression G. They are

$$G(t) \text{ has bounded variation, } -\infty < t < \infty. \tag{AI}$$

$$G(t) = 0, \qquad\qquad -\infty < t < 0. \tag{AII}$$

$$\int_{-\infty}^{\infty} |G(t)| dt \text{ exists.} \tag{AIII}$$

It may be shown [2] that under these conditions $G(t)$ may be expressed by the equation

$$G(t) = \frac{1}{2\pi i} \int_{I} J(i\omega) e^{i\omega t} d(i\omega), \tag{5}$$

where

$$J(i\omega) = \int_{-\infty}^{\infty} G(t) e^{-i\omega t} dt. \tag{6}$$

These expressions may be taken to define J. The function may, however, be obtained directly from computations or measurements; in the latter case the function is not defined for negative values of ω. It must be defined as follows to be consistent with the definition in (6):

$$J(-i\omega) = \text{complex conjugate of } J(i\omega). \tag{7}$$

While the final results will be expressed in terms of $AJ(i\omega)$ it will be convenient for the purpose of the intervening mathematics to define an auxiliary and closely related function

$$w(z) = \frac{1}{2\pi i} \int_{I} \frac{AJ(i\omega)}{i\omega - z} d(i\omega), \qquad 0 < x < \infty, \tag{8}$$

where

$$z = x + iy \tag{9}$$

and where x and y are real. Further, we shall define

$$w(iy) = \lim_{z \to 0} w(z). \tag{10}$$

---

[2] See Appendix II for fuller discussion.

The function will not be defined for $x < 0$ nor for $|z| = \infty$. As defined it is analytic[3] for $0 < x < \infty$ and at least continuous for $x = 0$.

The following restrictions on the network may be deduced:

$$\lim_{y \to \infty} y\,|J(iy)| \text{ exists.} \tag{BI}$$

$$J(iy) \text{ is continuous.} \tag{BII}$$

$$w(iy) = A\,J(iy). \tag{BIII}$$

Equation (5) may now be written

$$AG(t) = \frac{1}{2\pi i}\int_I w(z)e^{zt}dz = \frac{1}{2\pi i}\int_{s+} w(z)e^{zt}dz. \tag{11}$$

From a physical standpoint these restrictions are not of consequence. Any network made up of positive resistances, conductances, inductances, and capacitances meets them. Restriction (AII) says that the response must not precede the cause and is obviously fulfilled physically. Restriction (AIII) is fulfilled if the response dies out at least exponentially, which is also assured. Restriction (AI) says that the transmission must fall off with frequency. Physically there are always enough distributed constants present to insure this. This effect will be illustrated in example 8 below. Every physical network falls off in transmission sooner or later and it is ample for our purposes if it begins to fall off, say, at optical frequencies. We may say then that the reasoning applies to all linear networks which occur in nature. It also applies to other linear networks which are not physically producible but which may be specified mathematically. See example 7 below.

A temporary wave $f_0(t)$ is to be introduced into the system and an investigation will be made of whether the resultant disturbance in the system dies out. It has associated with it a function $F(z)$ defined by

$$f_0(t) = \frac{1}{2\pi i}\int_I F(z)e^{zt}dz = \frac{1}{2\pi i}\int_{s+} F(z)e^{zt}dz. \tag{12}$$

$F(z)$ and $f_0(t)$ are to be made subject to the same restrictions as $w(z)$ and $G(t)$ respectively.

Derivation of a Series for the Total Current

Let the amplifier be linear and of infinite power-carrying capacity. Let the output be connected to the input in such a way that the

[3] W. F. Osgood, "Lehrbuch der Funktionentheorie," 5th ed., Kap. 7, § 1, Hauptsatz. For definition of "analytic" see Kap. 6, § 5.

amplification ratio for one round trip is equal to the complex quantity AJ, where A is a function of the gain only and J is a function of ω only, being defined for all values of frequency from 0 to ∞.

Let the disturbing wave $f_0(t)$ be applied anywhere in the circuit. We have

$$f_0(t) = \frac{1}{2\pi} \int_{-\infty}^{+\infty} F(i\omega) e^{i\omega t} d\omega \tag{13}$$

or

$$f_0(t) = \frac{1}{2\pi i} \int_{s+} F(z) e^{zt} dz. \tag{13'}$$

The wave traverses the circuit and on completing the first trip it becomes

$$f_1(t) = \frac{1}{2\pi} \int_{-\infty}^{\infty} w(i\omega) F(i\omega) e^{i\omega t} d\omega \tag{14}$$

$$= \frac{1}{2\pi i} \int_{s+} w(z) F(z) e^{zt} dz. \tag{14'}$$

After traversing the circuit a second time it becomes

$$f_2(t) = \frac{1}{2\pi i} \int_{s+} F w^2 e^{zt} dz, \tag{15}$$

and after traversing the circuit n times

$$f_n(t) = \frac{1}{2\pi i} \int_{s+} F w^n e^{zt} dz. \tag{16}$$

Adding the voltage of the original impulse and the first n round trips we have a total of

$$s_n(t) = \sum_{k=0}^{n} f_k(t) = \frac{1}{2\pi i} \int_{s+} F(1 + w + \cdots w^n) e^{zt} dz. \tag{17}$$

The total voltage at the point in question at the time t is given by the limiting value which (17) approaches as n is increased indefinitely [4]

$$s(t) = \sum_{k=0}^{\infty} f_k(t) = \lim_{n \to \infty} \frac{1}{2\pi i} \int_{s+} S_n(z) e^{zt} dz, \tag{18}$$

where

$$S_n = F + Fw + Fw^2 + \cdots Fw^n = \frac{F(1 - w^{n+1})}{1 - w}. \tag{19}$$

[4] Mr. Carson has called my attention to the fact that this series can also be derived from Theorem IX, p. 49, of his Electric Circuit Theory. Whereas the present derivation is analogous to the theory expressed in equations (a)–(e) above, the alternative derivation would be analogous to that in equations (f)–(h).

REGENERATION THEORY   133

Convergence of Series

We shall next prove that the limit $s(t)$ exists for all finite values of t. It may be stated as of incidental interest that the limit

$$\int_{s+} S_\infty(z)e^{izt}dz \tag{20}$$

does not necessarily exist although the limit $s(t)$ does. Choose M_0 and N such that

$$|f_0(\lambda)| \leq M_0. \qquad 0 \leq \lambda \leq t. \tag{21}$$

$$|G(t - \lambda)| \leq N. \quad 0 \leq \lambda \leq t. \tag{22}$$

We may write [5]

$$f_1(t) = \int_{-\infty}^{\infty} G(t - \lambda)f_0(\lambda)d\lambda. \tag{23}$$

$$|f_1(t)| \leq \int_0^t M_0 N d\lambda = M_0 N t. \tag{24}$$

$$f_2(t) = \int_{-\infty}^{\infty} G(t - \lambda)f_1(\lambda)d\lambda. \tag{25}$$

$$|f_2(t)| \leq \int_0^t M_0 N^2 t dt = M_0 N^2 t^2/2! \tag{26}$$

Similarly

$$|f_n(t)| \leq M_0 N^n t^n/n! \tag{27}$$

$$|s_n(t)| \leq M_0(1 + Nt + \cdots N^n t^n/n!). \tag{28}$$

It is shown in almost any text [6] dealing with the convergence of series that the series in parentheses converges to e^{Nt} as n increases indefinitely. Consequently, $s_n(t)$ converges absolutely as n increases indefinitely.

Relation Between $s(t)$ and w

Next consider what happens to $s(t)$ as t increases. As t increases indefinitely $s(t)$ may converge to zero, indicating a condition of stability, or it may go beyond any value however large, indicating a runaway condition. The question which presents itself is: *Referring to (18) and (19), what properties of $w(z)$ and further what properties of $A J(i\omega)$ determine whether $s(t)$ converges to zero or diverges as t increases*

[5] G. A. Campbell, "Fourier Integral," *B. S. T. J.*, Oct. 1928, Pair 202.
[6] E.g., Whittaker and Watson, "Modern Analysis," 2d ed., p. 531.

indefinitely? From (18) and (19)

$$s(t) = \lim_{n \to \infty} \frac{1}{2\pi i} \int_{s+} F\left(\frac{1}{1-w} - \frac{w^{n+1}}{1-w}\right) e^{zt} dz. \qquad (29)$$

We may write

$$s(t) = \frac{1}{2\pi i} \int_{s+} [F/(1-w)] e^{zt} dz - \lim_{n \to \infty} \frac{1}{2\pi i} \int_{s+} [Fw^{n+1}/(1-w)] e^{zt} dz \qquad (30)$$

provided these functions exist. Let them be called $q_0(t)$ and $\lim_{n \to \infty} q_n(t)$ respectively. Then

$$q_n(t) = \int_{-\infty}^{\infty} q_0(t - \lambda) \phi(\lambda) d\lambda. \qquad (31)$$

where

$$\phi(\lambda) = \frac{1}{2\pi i} \int_{s+} w^{n+1} e^{z\lambda} dz. \qquad (32)$$

By the methods used under the discussion of convergence above it can then be shown that this expression exists and approaches zero as n increases indefinitely provided $q_0(t)$ exists and is equal to zero for $t < 0$. Equation (29) may therefore be written, subject to these conditions

$$s(t) = \frac{1}{2\pi i} \int_{s+} [F/(1-w)] e^{zt} dz. \qquad (33)$$

In the first place the integral is zero for negative values of t because the integrand approaches zero faster than the path of integration increases. Moreover,

$$\int_I [F/(1-w)] e^{zt} dz \qquad (34)$$

exists for all values of t and approaches zero for large values of t if $1 - w$ does not equal zero on the imaginary axis. Moreover, the integral

$$\int_C [F/(1-w)] e^{zt} dz \qquad (35)$$

exists because

1. Since F and w are both analytic within the curve the integrand does not have any essential singularity there,
2. The poles, if any, lie within a finite distance of the origin because $w \to 0$ as $|z|$ increases, and
3. These two statements insure that the total number of poles is finite.

We shall next evaluate the integral for a very large value of t. It will suffice to take the C integral since the I integral approaches zero. Assume originally that $1 - w$ does not have a root on the imaginary axis and that $F(z)$ has the special value $w'(z)$. The integral may be written

$$\frac{1}{2\pi i} \int_C [w'/(1-w)]e^{zt}dz. \tag{36}$$

Changing variables it becomes

$$\frac{1}{2\pi i} \int_D [1/(1-w)]e^{zt}dw, \tag{37}$$

where z is a function of w and D is the curve in the w plane which corresponds to the curve C in the z plane. More specifically the imaginary axis becomes the locus $x = 0$ and the semicircle becomes a small curve which spirals around the origin. See Fig. 2. The function

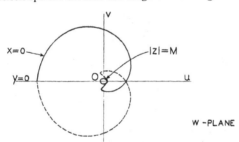

Fig. 2—Representative paths of integration in the w-plane corresponding to paths in Fig. 1.

z and, therefore, the integrand is, in general, multivalued and the curve of integration must be considered as carried out over the appropriate Riemann surface.[7]

Now let the path of integration shrink, taking care that it does not shrink across the pole at $w = 1$ and initially that it does not shrink across such branch points as interfere with its passage, if any. This shrinking does not alter the integral [8] because the integrand is analytic at all other points. At branch points which interfere with the passage of the path the branches stopped may be severed, transposed and connected in such a way that the shrinking may be continued past the branch point. This can be done without altering the value of the integral. Thus the curve can be shrunk until it becomes one or more very small circles surrounding the pole. The value of the total integral

[7] Osgood, loc. cit., Kap. 8.
[8] Osgood, loc. cit., Kap. 7, § 3, Satz 1.

(for very large values of t) is by the method of residues [9]

$$\sum_{j=1}^{n} r_j e^{z_j t},$$ (38)

where z_j ($j = 1, 2 \cdots n$) is a root of $1 - w = 0$ and r_j is its order. The real part of z_j is positive because the curve in Fig. 1 encloses points with $x > 0$ only. The system is therefore stable or unstable according to whether

$$\sum_{j=1}^{n} r_j$$

is equal to zero or not. But the latter expression is seen from the procedure just gone through to equal the number of times that the locus $x = 0$ encircles the point $w = 1$.

If F does not equal w' the calculation is somewhat longer but not essentially different. The integral then equals

$$\sum_{j=1}^{n} \frac{F(z_j)}{w(z_j)} e^{z_j t}$$ (39)

if all the roots of $1 - w = 0$ are distinct. If the roots are not distinct the expression becomes

$$\sum_{j=1}^{n} \sum_{k=1}^{r_j} A_{jk} t^{k-1} e^{z_j t},$$ (40)

where A_{jr_j}, at least, is finite and different from zero for general values of F. It appears then that unless F is specially chosen the result is essentially the same as for $F = w'$. The circuit is stable if the point lies wholly outside the locus $x = 0$. It is unstable if the point is within the curve. It can also be shown that if the point is on the curve conditions are unstable. We may now enunciate the following

Rule: Plot plus and minus the imaginary part of $AJ(i\omega)$ against the real part for all frequencies from 0 to ∞. If the point $1 + i0$ lies completely outside this curve the system is stable; if not it is unstable.

In case of doubt as to whether a point is inside or outside the curve the following criterion may be used: Draw a line from the point $(u = 1, v = 0)$ to the point $z = -i\infty$. Keep one end of the line fixed at $(u = 1, v = 0)$ and let the other end describe the curve from $z = -i\infty$ to $z = i\infty$, these two points being the same in the w plane. If the net angle through which the line turns is zero the point $(u = 1, v = 0)$ is on the outside, otherwise it is on the inside.

If AJ be written $|AJ|(\cos \theta + i \sin \theta)$ and if the angle always

---

[9] Osgood, loc. cit., Kap. 7, § 11, Satz 1.

changes in the same direction with increasing ω, where ω is real, the rule can be stated as follows: The system is stable or unstable according to whether or not a real frequency exists for which the feed-back ratio is real and equal to or greater than unity.

In case $d\theta/d\omega$ changes sign we may have the case illustrated in Figs. 3 and 4. In these cases there are frequencies for which w is real and

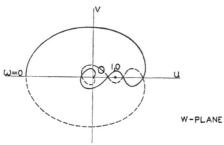

Fig. 3—Illustrating case where amplifying ratio is real and greater than unity for two frequencies, but where nevertheless the path of integration does not include the point 1, 0.

greater than 1. On the other hand, the point (1, 0) is outside of the locus $x = 0$ and, therefore, according to the rule there is a stable condition.

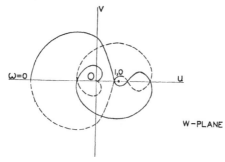

Fig. 4—Illustrating case where amplifying ratio is real and greater than unity for two frequencies, but where nevertheless the path of integration does not include the point 1, 0.

If networks of this type were used we should have the following interesting sequence of events: For low values of A the system is in a stable condition. Then as the gain is increased gradually, the system becomes unstable. Then as the gain is increased gradually still further, the system again becomes stable. As the gain is still further increased the system may again become unstable.

138   BELL SYSTEM TECHNICAL JOURNAL

Examples

The following examples are intended to give a more detailed picture of certain rather simple special cases. They serve to illustrate the previous discussion. In all the cases F is taken equal to AJ so that f_0 is equal to AG. This simplifies the discussion but does not detract from the illustrative value.

1. Let the network be pure resistance except for the distortionless amplifier and a single bridged condenser, and let the amplifier be such that there is no reversal. We have

$$AJ(i\omega) = \frac{B}{\alpha + i\omega}, \qquad (41)$$

where A and α are real positive constants. In (18) [10]

$$f_n = \frac{1}{2\pi i} \int_I A^{n+1} J^{n+1}(i\omega) e^{i\omega t} d i\omega \qquad (42)$$
$$= Be^{-\alpha t}(B^n t^n / n!).$$

$$s(t) = Be^{-\alpha t}(1 + Bt + B^2 t^2 / 2! + \cdots). \qquad (43)$$

The successive terms f_0, f_1, etc., represent the impressed wave and the successive round trips. The whole series is the total current.

It is suggested that the reader should sketch the first few terms graphically for $B = \alpha$, and sketch the admittance diagrams for $B < \alpha$, and $B > \alpha$.

The expression in parentheses equals e^{Bt} and

$$s(t) = Be^{(B-\alpha)t}. \qquad (44)$$

This expression will be seen to converge to 0 as t increases or fail to do so according to whether $B < \alpha$ or $B \geq \alpha$. This will be found to check the rule as applied to the admittance diagram.

2. Let the network be as in 1 except that the amplifier is so arranged that there is a reversal. Then

$$AJ(i\omega) = \frac{-B}{\alpha + i\omega}. \qquad (45)$$

$$f_n = (-1)^{n+1} Be^{-\alpha t}(B^n t^n / n!). \qquad (46)$$

The solution is the same as in 1 except that every other term in the series has its sign reversed:

$$s(t) = -Be^{-\alpha t}(1 - Bt + B^2 t^2 / 2! + \cdots)$$
$$= -Be^{(-\alpha - B)t}. \qquad (47)$$

[10] Campbell, loc. cit. Pair 105.

This converges to 0 as t increases regardless of how great B may be taken. If the admittance diagram is drawn this is again found to check the rule.

3. Let the network be as in 1 except that there are two separated condensers bridged across resistance circuits. Then

$$A J(i\omega) = \frac{B^2}{(\alpha + i\omega)^2}. \tag{48}$$

The solution for $s(t)$ is obtained most simply by taking every other term in the series obtained in 1.

$$s(t) = Be^{-\alpha t}(Bt + B^3 t^3/3! + \cdots)$$
$$= Be^{-\alpha t} \sinh Bt. \tag{49}$$

4. Let the network be as in 3 except that there is a reversal. Then

$$A J(i\omega) = \frac{- B^2}{(\alpha + i\omega)^2}. \tag{50}$$

The solution is obtained most directly by reversing the sign of every other term in the series obtained in 3.

$$s(t) = - Be^{-\alpha t}(Bt - B^3 t^3/3! + \cdots)$$
$$= - Be^{-\alpha t} \sin Bt. \tag{51}$$

This is a most instructive example. An approximate diagram has been made in Fig. 5, which shows that as the gain is increased the

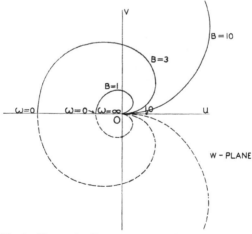

Fig. 5--Illustrating Example 4, with three values for B.

feed-back ratio may be made arbitrarily great and the angle arbitrarily small without the condition being unstable.   This agrees with the expression just obtained, which shows that the only effect of increasing the gain is to increase the frequency of the resulting transient.

5. Let the conditions be as in 1 and 3 except for the fact that four separated condensers are used.   Then

$$AJ(i\omega) = \frac{B^4}{(\alpha + i\omega)^4}.\tag{52}$$

The solution is most readily obtained by selecting every fourth term in the series obtained in 1.

$$s(t) = Be^{-\alpha t}(B^3t^3/3! + B^7t^7/7! + \cdots)$$
$$= \tfrac{1}{2}Be^{-\alpha t}\,(\sinh Bt - \sin Bt).\tag{53}$$

This indicates a condition of instability when $B \geq \alpha$, agreeing with the result deducible from the admittance diagram.

6. Let the conditions be as in 5 except that there is a reversal. Then

$$Y = \frac{-B^4}{(\alpha + i\omega)^4}.\tag{54}$$

The solution is most readily obtained by changing the sign of every other term in the series obtained in 5.

$$s(t) = Be^{-\alpha t}(- B^3t^3/3! + B^7t^7/7! - \cdots).\tag{55}$$

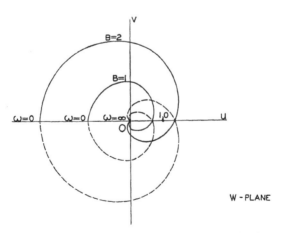

Fig. 6—Illustrating Example 6, with two values for B.

For large values of t this approaches

$$s(t) = -\tfrac{1}{2}Be^{(B/\sqrt{2}-\alpha)t}\sin{(Bt/\sqrt{2}-\pi/4)}. \qquad (56)$$

This example is interesting because it shows a case of instability although there is a reversal.   Fig. 6 shows the admittance diagram for

$B\sqrt{2}-\alpha<0$ and for $B\sqrt{2}-\alpha>0$.

7. Let

$$AG(t) = f_0(t) = A(1-t), \qquad 0 \le t \le 1. \qquad (57)$$

$$AG(t) = f_0(t) = 0, \qquad -\infty < t < 0, \qquad 1 < t < \infty. \qquad (57')$$

We have

$$AJ(i\omega) = A\int_0^1 (1-t)e^{-i\omega t}dt$$

$$= A\left(\frac{1-e^{-i\omega}}{\omega^2}+\frac{1}{i\omega}\right). \qquad (58)$$

Fig. 7 is a plot of this case for $A = 1$.

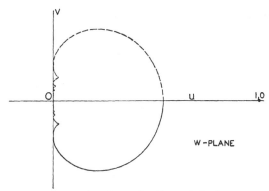

Fig. 7—Illustrating Example 7.

8. Let

$$AJ(i\omega) = \frac{A(1+i\omega)}{(1+i2\omega)}. \qquad (59)$$

This is plotted on Fig. 8 for $A = 3$.   It will be seen that the point 1 lies outside of the locus and for that reason we should expect that the system would be stable.   We should expect from inspecting the diagram that the system would be stable for $A < 1$ and $A > 2$ and that it would be unstable for $1 \le A \le 2$.   We have overlooked one fact, however; the expression for $AJ(i\omega)$ does not approach zero as ω

increases indefinitely. Therefore, it does not come within restriction (BI) and consequently the reasoning leading up to the rule does not apply.

The admittance in question can be made up by bridging a capacity in series with a resistance across a resistance line. This admittance

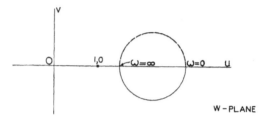

Fig. 8—Illustrating Example 8, without distributed constants.

obviously does not approach zero as the frequency increases. In any actual network there would, however, be a small amount of distributed capacity which, as the frequency is increased indefinitely, would cause the transmission through the network to approach zero. This is shown graphically in Fig. 9. The effect of the distributed capacity is

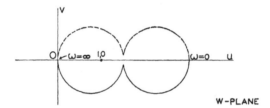

Fig. 9—Illustrating Example 8, with distributed constants.

essentially to cut a corridor from the circle in Fig. 8 to the origin, which insures that the point lies inside the locus.

APPENDIX I

Alternative Procedure

In some cases $AJ(i\omega)$ may be given as an analytic expression in $(i\omega)$. In that case the analytic expression may be used to define w for all values of z for which it exists. If the value for $AJ(i\omega)$ satisfies all the restrictions the value thus defined equals the w defined above for $0 \leq x < \infty$ only. For $-\infty < x < 0$ it equals the analytic continuation of the function w defined above. If there are no essential

REGENERATION THEORY 143

singularities anywhere including at ∞, the integral in (33) may be evaluated by the theory of residues by completing the path of integration so that all the poles of the integrand are included. We then have

$$s(t) = \sum_{j=1}^{j=n} \sum_{k=1}^{r_j} A_{jk} t^{k-1} e^{z_j t}. \tag{60}$$

If the network is made up of a finite number of lumped constants there is no essential singularity and the preceding expression converges because it has only a finite number of terms. In other cases there is an infinite number of terms, but the expression may still be expected to converge, at least, in the usual case. Then the system is stable if all the roots of $1 - w = 0$ have $x < 0$. If some of the roots have $x \geq 0$ the system is unstable.

The calculation then divides into three parts:

1. The recognition that the impedance function is $1 - w$.[11]

2. The determination of whether the impedance function has zeros for which $x \geq 0$.[12]

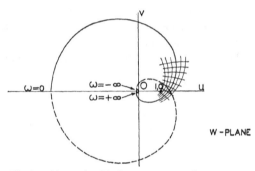

Fig. 10—Network of loci $x = $ const., and $y = $ const.

3. A deduction of a rule for determining whether there are roots for which $x \geq 0$. The actual solution of the equation is usually too laborious.

To proceed with the third step, plot the locus $x = 0$ in the w plane, i.e., plot the imaginary part of w against the real part for all the values of y, $-\infty < y < \infty$. See Fig. 10. Other loci representing

$$x = \text{const.} \tag{61}$$

and

$$y = \text{const.} \tag{62}$$

[11] Cf. H. W. Nichols, *Phys. Rev.*, vol. 10, pp. 171–193, 1917.
[12] Cf. Thompson and Tait, "Natural Philosophy," vol. I, § 344.

may be considered and are indicated by the network shown in the figure in fine lines. On one side of the curve x is positive and on the other it is negative. Consider the equation

$$w(z) - 1 = 0$$

and what happens to it as A increases from a very small to a very large value. At first the locus $x = 0$ lies wholly to the left of the point. For this case the roots must have $x < 0$. As A increases there may come a time when the curve or successive convolutions of it will sweep over the point $w = 1$. For every such crossing at least one of the roots changes the sign of its x. We conclude that if the point $w = 1$ lies inside the curve the system is unstable. It is now possible to enunciate the rule as given in the main part of the paper but there deduced with what appears to be a more general method.

Appendix II

Discussion of Restrictions

The purpose of this appendix is to discuss more fully the restrictions which are placed on the functions defining the network. A full discussion in the main text would have interrupted the main argument too much.

Define an additional function

$$n(z) = \frac{1}{2\pi i} \int_I \frac{A J(i\lambda)}{i\lambda - z} d(i\lambda), \qquad -\infty < x < 0. \tag{63}$$

$$n(iy) = \lim_{x \to 0} n(z).$$

This definition is similar to that for $w(z)$ given previously. It is shown in the theorem [13] referred to that these functions are analytic for $x \neq 0$ if $A J(i\omega)$ is continuous. We have not proved, as yet, that the restrictions placed on $G(t)$ necessarily imply that $J(i\omega)$ is continuous. For the time being we shall assume that $J(i\omega)$ may have finite discontinuities. The theorem need not be restricted to the case where $J(i\omega)$ is continuous. From an examination of the second proof it will be seen to be sufficient that $\int_I J(i\omega)d(i\omega)$ exist. Moreover, that proof can be slightly modified to include all cases where conditions (AI)–(AIII) are satisfied.

[13] Osgood, loc. cit.

For, from the equation at top of page 298 [13]

$$\left| \frac{w(z_0 - \Delta z) - w(z_0)}{\Delta z} - \frac{1}{2\pi i} \int_I \frac{A J(i\lambda)}{(i\lambda - z_0)^2} d(i\lambda) \right|$$

$$\leq |\Delta z| \left| \frac{1}{2\pi i} \int_I \frac{A J(i\lambda) d(i\lambda)}{(i\lambda - z_0 - \Delta z)(i\lambda - z_0)^2} \right|, \qquad x_0 > 0. \quad (64)$$

It is required to show that the integral exists.   Now

$$\int_I \frac{A J(i\lambda) d(i\lambda)}{(i\lambda - z_0 - \Delta z)(i\lambda - z_0)^2}$$

$$= \int_I \frac{A J(i\lambda) d(i\lambda)}{(i\lambda - z_0)^3} \left(1 + \frac{\Delta z}{i\lambda - z_0} + \frac{\Delta z^2}{i\lambda - z_0} + \text{etc.} \right) \quad (65)$$

if Δz is taken small enough so the series converges.   It will be sufficient to confine attention to the first term.   Divide the path of integration into three parts,

$$-\infty < \lambda < -|z_0| - 1, \qquad -|z_0| - 1 < \lambda < |z_0| + 1, \qquad |z_0| + 1 < \lambda < \infty.$$

In the middle part the integral exists because both the integrand and the range of integration are finite.   In the other ranges the integral exists if the integrand falls off sufficiently rapidly with increasing λ. It is sufficient for this purpose that condition (BI) be satisfied.   The same proof applies to $n(z)$.

Next, consider $\lim_{x \to 0} w(z) = w(iy)$.   If iy is a point where $J(iy)$ is continuous, a straightforward calculation yields

$$w(iy) = A J(iy)/2 + P(iy). \qquad (66a)$$

Likewise,

$$n(iy) = - A J(iy)/2 + P(iy) \qquad (66b)$$

where $P(iy)$ is the principal value [14] of the integral

$$\frac{1}{2\pi i} \int_I \frac{A J(i\lambda)}{i\lambda - iy} d(i\lambda).$$

Subtracting

$$w(iy) - n(iy) = A J(iy) \qquad (67)$$

If (iy) is a point of discontinuity of $J(iy)$

$$|w| \text{ and } |n| \text{ increase indefinitely as } x \to 0. \qquad (68)$$

Next, evaluate the integral

$$\frac{1}{2\pi i} \int_{x+I} w(z) e^{zt} dz,$$

[14] E. W. Hobson, "Functions of a Real Variable," vol. I, 3d edition, § 352.

146   *BELL SYSTEM TECHNICAL JOURNAL*

where the path of integration is from $x - i\infty$ to $x + i\infty$ along the line $x = $ const. On account of the analytic nature of the integrand this integral is independent of x (for $x > 0$). It may be written then

$$\lim_{x \to 0} \frac{1}{2\pi i} \int_{x+I} w(z) e^{zt} dz = \lim_{x \to 0} \frac{1}{2\pi i} \int_{x+I} \frac{1}{2\pi i} \int_I \frac{A J(i\lambda)}{i\lambda - z} e^{zt} d(i\lambda) dz$$

$$= \lim_{x \to 0} \frac{1}{2\pi i} \int_{x+I} \frac{1}{2\pi i} \lim_{M \to \infty} \left[\int_{-iM}^{iy-i\delta} + \int_{iy-i\delta}^{iy+i\delta} + \int_{iy+i\delta}^{iM} \right] \frac{A J(i\lambda)}{i\lambda - z} e^{zt} d(i\lambda) dz$$

$$= \lim_{x \to 0} \left[\frac{1}{2\pi i} \int_{x+I} \frac{1}{2\pi i} \int_{iy-i\delta}^{iy+i\delta} \frac{A J(i\lambda)}{i\lambda - z} e^{zt} d(i\lambda) dz + Q(t, \delta) \right], \quad x > 0, \quad (69)$$

where δ is real and positive. The function Q defined by this equation exists for all values of t and for all values of δ. Similarly,

$$\lim_{x \to 0} \frac{1}{2\pi i} \int_{x+I} n(z) e^{zt} dz$$

$$= \left[\lim_{x \to 0} \frac{1}{2\pi i} \int_{x+I} \frac{1}{2\pi i} \int_{iy+i\delta}^{iy+i\delta} \frac{A J(i\lambda)}{i\lambda - z} e^{zt} d(i\lambda) dz + Q(t, \delta) \right], \quad x < 0, \quad (70)$$

Subtracting and dropping the limit designations

$$\frac{1}{2\pi i} \int_{x+I} w(z) e^{zt} dz - \frac{1}{2\pi i} \int_{x+I} n(z) e^{zt} dz = \frac{1}{2\pi i} \int_I A J(i\lambda) e^{i\lambda t} d(i\lambda). \quad (71)$$

The first integral is zero for $t < 0$ as can be seen by taking x sufficiently large. Likewise, the second is equal to zero for $t > 0$. Therefore,

$$\frac{1}{2\pi i} \int_{x+I} w(z) e^{zt} dz = \frac{1}{2\pi i} \int_I A J(i\omega) e^{i\omega t} d(i\omega) = AG(t), \quad 0 < t < \infty \quad (72)$$

$$-\frac{1}{2\pi i} \int_{x+I} n(z) e^{zt} dz$$

$$= \frac{1}{2\pi i} \int_I A J(i\omega) e^{i\omega t} d(i\omega) = AG(t) - \infty < t < 0. \quad (73)$$

We may now conclude that

$$\int_I n(iy) e^{iyt} d(iy) = 0, \quad -\infty < t < \infty \quad (74)$$

provided

$$G(t) = 0, \quad -\infty < t < 0. \quad (AII)$$

But (74) is equivalent to

$$n(z) = 0, \quad (74')$$

which taken with (67) gives

$$w(iy) = A J(iy). \tag{BIII}$$

(BIII) is, therefore, a necessary consequence of (AII). (74') taken with (68) shows that

$$J(iy) \text{ is continuous.} \tag{BII}$$

It may be shown [15] that (BI) is a consequence of (AI). Consequently all the B conditions are deducible from the A conditions.

Conversely, it may be inquired whether the A conditions are deducible from the B conditions. This is of interest if $A J(i\omega)$ is given and is known to satisfy the B conditions, whereas nothing is known about G.

Condition AII is a consequence of BIII as may be seen from (67) and (74). On the other hand AI and AIII cannot be inferred from the B conditions. It can be shown by examining (5), however, that if the slightly more severe condition

$$\lim_{y \to \infty} y^\gamma J(iy) \text{ exists}, \qquad (\gamma > 1), \tag{BIa}$$

is satisfied then

$$G(t) \text{ exists}, \qquad -\infty < t < \infty, \tag{AIa}$$

which, together with AII, insures the validity of the reasoning.

It remains to show that the measured value of $J(i\omega)$ is equal to that defined by (6). The measurement consists essentially in applying a sinusoidal wave and determining the response after a long period. Let the impressed wave be

$$E = \text{real part of } e^{i\omega t}, \qquad t \geq 0. \tag{75}$$
$$E = 0, \qquad t < 0. \tag{75'}$$

The response is

$$\text{real part of } \int_0^t A G(\lambda) e^{i\omega(t-\lambda)} d\lambda$$

$$= \text{real part of } A e^{i\omega t} \int_0^t G(\lambda) e^{-i\omega\lambda} d\lambda. \tag{76}$$

For large values of t this approaches

$$\text{real part of } A e^{i\omega t} J(i\omega). \tag{77}$$

Consequently, the measurements yield the value $A J(i\omega)$.

[15] See Hobson, loc. cit., vol. II, 2d edition, § 335. It will be apparent that K depends on the total variation but is independent of the limits of integration.

FEEDBACK—THE HISTORY OF AN IDEA*

by H. W. Bode

H. W. BODE, now a vice president of Bell Telephone Laboratories, is notable for his many pioneering efforts in the theory of automatic control. In this review article he illumines some of the contributions of theoreticians in the communications area. Several of the basic ideas of Black, Nyquist, MacMillan, and MacCall are delineated against a background of the development of transcontinental telephony. The discussion ranges from negative feedback amplifiers to adaptive control devices.

* From *Proceedings of the Symposium on Active Networks and Feedback Systems,* Polytechnic Institute of Brooklyn, Polytechnic Press, 1960.

FEEDBACK - THE HISTORY OF AN IDEA

H. W. Bode
Bell Telephone Laboratories, Inc., Whippany, N. J.

The modern field of control theory as it is now understood is a marriage of two originally quite separate technical areas. One area is the classic field of regulators, as exemplified by the household thermostat or the Watt centrifugal governor. The other is the comparatively recent theory of negative feedback amplifiers, originally developed to meet the problems of long-distance telephony. The two fields were brought together by military servomechanism and computer problems which arose during the war.

The paper attempts to review this history, with special reference to the early development of the negative feedback amplifier. It closes with a brief assessment of the long-term implications of this marriage as seen from a present-day point of view.

Social philosophers say that this is above all an age of mobility. In past times, most boys grew up, lived and died close to the homes of their parents, but this is now the exception. I dare say that only a very small fraction of this audience came from less than several hundred miles away. Most of us, I am sure, have lived in a number of different towns, and probably have not lived in the towns in which we grew up since our childhood. Thus, many of you perhaps have had the experience that I had recently of revisiting, after a long absence, a town that I had known as a child. This was my grandparents' town in rural Illinois. I knew it in the days when, in those parts at least, the standard transportation was still the horse and buggy, and it took a pretty strong horse and a pretty light buggy to get through those muddy roads sometimes. I had very pleasant recollections of a quiet town of the New England type, full of fine old houses and elm-shaded streets. When I had an opportunity to go back there a few years ago, I returned full of nostalgia and sentiment.

It was a very disconcerting experience. I found that my recollections weren't at all right. I couldn't find my way around the town at all. The fine old elms had long since been cut down. Many of the old houses had been pulled down and the ones that weren't were pretty shabby. There was a lot of impressive new construction, but the town's layout, geography, and atmosphere of hustle were very different from anything I remembered.

I tried to talk a little to the hotel clerk about my impressions,

Presented at the *Symposium on Active Networks and Feedback Systems,*
Polytechnic Institute of Brooklyn, April 19-21, 1960

2   ACTIVE NETWORKS AND FEEDBACK SYSTEMS

including a few misgivings I had developed. The town when I knew it
was centered very pleasantly on the village square and had a well defined structure of its own. Now it seemed to straggle for miles along
the highway, a mish-mash of hot dog stands, small real estate developments, gas stations, and whatnot. But the clerk couldn't have
been less interested. He'd been there five or six years now, it was
his town, he was an old inhabitant, he liked it the way it was and was
a little hostile to any kind of criticism. Some of my friends tell me
that, at that, I was lucky. Their hometowns have simply disappeared
--swallowed up by the expansion of such modern megalopolises as
Detroit or Los Angeles.

I was reminded of all that when I started to prepare for this
paper. In inviting me to give a paper, Professor Carlin knew, of
course, that I hadn't worked in the feedback field for nearly 20 years.
My book on the subject appeared after the war, but it was actually
written before the war began. Professor Carlin suggested, in particular, that an appropriate talk might be devoted largely to some
reminiscences of the early days of the field, perhaps including
also an indication of the way in which the seeds planted then have developed into modern control theory, and a brief critique of the field
as it now stands.

I approached this task full, again, of nostalgia and sentiment.
As I looked over some of the newer literature, however, I found myself again less at home than I had expected. This "town" also has
grown beyond my recollection. I no longer feel that I know it well
enough to treat it in any comprehensive way. What I would like to do
this morning, consequently, is only to trace one or two threads of the
feedback story which I happen to know most about, and which seem to
have a certain continuity. With your permission, I would also like to
express a few misgivings I have concerning the present logical structure of the control field, which, like my grandparents' hometown,
seems to become more sprawling as it grows, but I will not attempt
any broad appraisal of the field.

To begin, then, we may recall that the modern feedback and
control field is the fusion of what were originally two quite different
technological areas. One parent area is represented by the typical
mechanical regulator or elementary control circuit, such as the
household thermostat. This is a very vast as well as a very old field.
One reference I read credited a certain Papin with the first known invention in this area, in 1680. The invention consisted of a weighted
saucepan lid to provide pressure cooking to a definite limit of pressure. One suspects that this particular flash of genius may well have
occurred, in informal ways, to many people before Papin, since even
our stone age ancestors had ample background in the use of rocks and
other heavy objects to hold things down. However, Papin's particular
inspiration did lead to a variety of steam safety valves and similar
devices still used in modern technology. The Watt centrifugal governor

FEEDBACK   3

is, of course, another and more famous invention for the control of steam engines. Other examples of such regulating systems are furnished by escapements for clocks, steering engines and roll control systems for ships, etc.

The second parent area is much more recent. It resulted essentially from the efforts of communication engineers, and particularly telephone engineers, to provide extremely high quality amplifiers. These stemmed, as I will show in a moment, from the need to provide satisfactory long distance telephone service in a reasonably economical way. The same work also underlies much of the systematic theory of active networks which we now have.

The two parent fields were originally quite different in character and emphasis. Thus, the typical regulator was essentially a "dc" device. The designer was interested in attaining a given operating point with fairly high precision, but within broad limits he was not much interested in the exact transient performance of the system in reaching the equilibrium point. The negative feedback devices used in telephony, on the other hand, were historically called "wave amplifiers, " because they had to follow and reproduce a randomly varying input signal wave with high fidelity. In other words, the circuit was always, in a sense, in the transient condition and its transient characteristics for a wide variety of excitations were all-important.

The two fields also differ radically in their mathematical flavor. The typical regulator system can frequently be described, in essentials, by differential equations of no more than perhaps the second, third or fourth order. On the other hand, the system is usually highly nonlinear, so that even at this level of complexity the difficulties of analysis may be very great. In contrast, the order of the set of differential equations describing the typical negative feedback amplifier used in telephony is likely to be very much greater. As a matter of idle curiosity, I once counted to find what the order of the set of equations in an amplifier I had just designed would have been, if I had worked with the differential equations directly. It turned out to be 55. On the other hand, the circuits are quite linear, even without counting the effects of feedback, so that the very powerful methods of linear analysis are available in studying them; one can, in fact, normally work with circuit equations in the integrated or "steady state" form.

The two fields were brought together very suddenly and very emphatically by the pressures of the war. The example of the servo control systems used to point the heavy guns on a warship is sufficient to tell why. If one ship is firing at another, they are likely to be maintaining nearly parallel courses, so that they will be moving rather slowly with respect to one another. Thus, the problem involved in servo control of the guns is essentially one of maintaining a modest lead angle which is nearly fixed. This is something like the typical regulator problem. On the other hand, if a ship is defending itself

4   ACTIVE NETWORKS AND FEEDBACK SYSTEMS

against an airplane attack the much greater speed and mobility of the
airplane means that the calculated lead angles must be both very large
and rapidly varying. Thus, what is required is something like a "wave
amplifier" which will follow a complex signal with fidelity.

In the course of the war there were, of course, many other similar situations, all reflecting the increased speed and maneuverability
of modern weapons. In addition, modern feedback theory turned out
to have numerous other applications in weaponry. For example, it
was very useful in the design of new kinds of sensing equipments, such
as radar. Because of the nulling properties of negative feedback
circuits, the theory also supplied an important tool in the design of
computers to solve complex implicit equations, such as frequently appear in fire control systems.

The parentage which stems from the needs of long distance telephony is the one that I am most familiar with and would like to turn to
now. Almost from its inception, the Bell System had in mind the de-
sirability of producing a nationwide network. It turned out to
be possible to link together cities as close as New York and Boston or
even New York and Chicago by using very heavy voice frequency conductors, open wire lines, and sufficient loading. However, the losses
in a transcontinental link were simply too great to be overcome without intermediate amplification. Various possible amplifiers were
tried. One, for example, was based on a conventional telephone receiver talking into a carbon button transmitter. But nothing worked
well enough until the vacuum tube amplifier was invented. This infant
device was grabbed enthusiastically, and with its help the continent
was finally spanned from New York to San Francisco in 1915. *

It is of interest to note what was required to do this job. The
conductors were No. 8, weighing almost half a ton per mile. At a
suitable voltage one could supply electric power to a good-size village
over such conductors. The circuits were loaded to a cut-off just
above 1,000 cycles. These measures reduced the transcontinental
loss to 60 db. With the help of six vacuum tube amplifiers, it was
possible to erase 42 db of this total, leaving a net loss of 18 db,
which was tolerable. Clearly, one did not think of large gains from
the vacuum tubes of those days.

As modest as this beginning seems by present standards, it
clearly promised great things. Vacuum tube repeaters, if they could
be made at all, would obviously be very cheap in comparison with
copper costs. All that seemed to be needed for a vast expansion in
long distance communications was an adequate advance in the state of
the art in vacuum tube repeaters.

At first, this development went swimmingly. With better vacuum tubes and repeater design, one could get along with smaller

*An experimental system involving only three amplifiers was actually put into
operation the preceding year.

FEEDBACK   5

conductors and less loading, and, by the application of carrier principles, even use the lines for more than one conversation. The long
distance telephone system grew correspondingly. Soon, however, the
course of events hit a serious roadblock. The problem can be understood from Table I. The entries in the table are merely representa-

SYSTEM	DATE	CHANNELS PER PAIR	LOSS IN DB (3000MI)	REPEATERS (3000MI)
1ST TRANSCONTINENTAL	1914	1	60	3-6
2ND TRANSCONTINENTAL	1923	1-4	150-400	6-20
OPEN WIRE CARRIER	1938	16	1000	40
CABLE CARRIER	1936	12	12,000	200
FIRST COAXIAL	1941	480	30,000	600

Table I: Toll repeater development

tive, but they are sufficient to suggest the basic problem. The first
line corresponds to the original transcontinental system which I have
just described. The second is taken to correspond to the opening of
the second transcontinental geographic route, through Texas to Los
Angeles. The date, 1923, for this system is also, by coincidence,
the year in which Harold S. Black began the research which led to the invention of the negative feedback amplifier a few years later. The remaining dates are those at which the corresponding illustrative systems were actually announced. However, the engineering development in each case was much earlier. With the possible exception of
the coaxial, they are systems which could have been envisaged at
least in broad terms even in 1923, but which could not be realized
until the negative feedback invention had been perfected.

The last two columns show what the roadblock was. As we are
more ambitious in our uses of carrier and repeater techniques, the
total loss in the line and consequently the total number of repeaters
must go up. But each repeater is a source of noise and distortion,
and the signal degradation produced by all these effects is directly
proportional to the total number of repeaters which must be traversed.
Most of you with hi-fi systems are no doubt proud of the quality of
your audio amplifiers, but I doubt whether many of you would care to
listen to the sound after the signal had gone in succession through
several dozen or several hundred even of your own fine amplifiers.
There is a "tyranny of numbers, " as my reliability friends say, which
makes it necessary for the individual components of the system to

6   ACTIVE NETWORKS AND FEEDBACK SYSTEMS

become qualitatively better as the system as a whole becomes quantitatively more ambitious. The systems envisaged in the third and succeeding lines of Table I simply required too many amplifiers in tandem to yield acceptable over-all quality until a substantial step forward in individual amplifier quality could be made.

The causes of distortion were of various sorts. They included power supply noises, variations in gain, and so on. The dominant problem, however, was the intermodulation due to the slight nonlinearity in the characteristics of the last tube. Various efforts were made to improve this situation, by the selection of tubes, by careful biasing, by the use of matched tubes in push-pull to provide compensating characteristics, and so on. Until Black's invention, however, nothing made a radical improvement in the situation.

Black actually made two inventions. The first[1] is now primarily a curiosity, although its principles are still of interest. It is shown in Fig. 1(a), which is intended to be illustrative rather than entirely accurate. The two boxes labeled "+" represent bi-conjugate devices, like three winding transformers or Wheatstone bridges, such

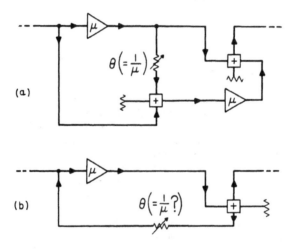

Fig.1 Early Black amplifiers

that transmission from any input can be obtained to either adjacent output but not to the facing output. The top amplifier is the principal one and transmits through the second bi-conjugate device to the output line. The second amplifier is a compensating device. In operation, the output signal from the principal amplifier is first reduced by the variable attenuator labeled θ and balanced against a sample from the input circuit. If we set $\theta = 1/\mu$, so that the loss (and phase) of the variable attenuator exactly equals the gain of the top amplifier, and

FEEDBACK   7

if the top amplifier is perfect in the sense that its output contains no
extraneous noise or modulation components, this balance is precise
and nothing is fed to the bottom amplifier. However, extraneous
noise or modulation from the top amplifier is not balanced out. Instead, it goes through the bottom amplifier and is combined, with reversed polarity, with the direct feed from the first amplifier, in the
second bi-conjugate circuit. Since the gain in the second amplifier
is just sufficient to balance the loss in the $1/\mu$ attenuator, these im-
perfections in the original amplifier output are cancelled by this process and do not appear in the final output line.

Since the lower amplifier normally carries only a small signal,
the fact that its characteristics may not be quite linear is not important. On the other hand, if it is made identical with the top amplifier,
it furnishes a valuable standby capability. As will readily be seen,
the transmission through the circuit is still correct even if the top
amplifier fails completely. While the circuit appears to have closed
loops, the balances in the two bi-conjugate circuits prevent any feedback currents from actually circulating through either amplifier.

Experimentally, it was possible to realize the advantages
promised by the circuit over narrow frequency bands. On the other
hand, it turned out to be difficult to maintain the balances in the bi-
conjugate circuits, or the balance between the gain of the μ circuits
and the loss of the $1/\mu$ attenuator, exactly enough over broad bands.
Thus, the additional cost and complexity of the compensating circuits
did not lead to a corresponding improvement in performance for typical applications, and Black felt compelled to look further for the ma-
jor advance he was seeking.

It is tantalizing to observe that we need a whole second amplifier identical with the primary one just to provide the compensation.
One would like to think that it should somehow be possible to use the
same amplifier twice. A naive approach to this problem is indicated
in the schematic of Fig. 1(b). Obviously, if we continue to set $\theta = 1/\mu$,
the input of the single amplifier in the new circuit will see the same
noise and distortion voltages which were applied to the compensating
amplifier in the original schematic, so that on a first trip through the
amplifier it appears to provide the same compensating effect at the
amplifier output. The new circuit does not work, however, because
the closed loop around the amplifier is no longer interrupted by the
balanced bridge. The compensating voltage, instead of doing its job
and expiring, goes round and round the loop. There is, in fact, no
way of setting θ so that exact compensation can be achieved.

Nevertheless, the configuration of Fig. 1(b) is still the prototype
of the finally successful negative feedback amplifier.[2] The essential
difference is merely that the loss θ instead of being equal to the gain of the
μ circuit is much smaller, so that there is a large net gain around the loop.
Black's invention lay in the recognition that while this leads to a mode
of operation quite different from the compensation scheme, it is still

8   ACTIVE NETWORKS AND FEEDBACK SYSTEMS

an effective way of reducing distortion. To obtain this insight, however, he had first to make a comprehensive study of the theoretical
aspects of distortion production in vacuum tube circuits. The actual
invention was made in 1927. As with so many basic inventions, the
final solution came to the inventor in a flash, but only after he had
been working hard on the problem for several years.

Although the broad implications of the negative feedback invention were rather quickly apparent, Black's original ideas required
support and confirmation in a number of directions before engineering
exploitation of the field could go forward expeditiously and with con-
fidence. For example, the invention relied basically upon amplifiers
with loop gains much in excess of unity. The fact that such structures
could exist, without instability, seemed doubtful in 1927 to many people. Obviously, in order to go further, we needed a practical criterion which would tell the design engineer what properties he should
seek to ensure a stable system.

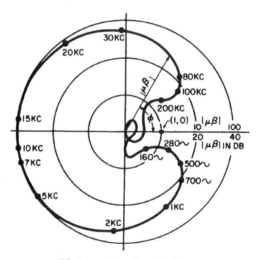

Fig.2 The Nyquist criterion

Fortunately for the application of the negative feedback amplifier, the required solution was quickly forthcoming in a classic analysis by Nyquist.[3] No doubt all of you are familiar with the Nyquist criterion, but to refresh your memories I have shown in Fig. 2 a plot of
the feedback characteristics of one of the early amplifiers. The critical point is 1, 0. As will be seen, the $\mu\beta$ characteristic plotted as
a vector quantity for all significant frequencies fails to circle the critical point. Thus, the amplifier should be stable, in agreement with
the result found experimentally when the amplifier was built.

Two other supporting elements in this area need brief mention.

FEEDBACK   9

One is a research by Peterson, Kreer and Ware[4] to demonstrate experimentally that the Nyquist criterion also holds even in the so-called
"conditional" case in which the $\mu\beta$ characteristic overlaps the axis
of abscissae before finally turning past the critical point. Curve (c)
in Fig. 3 shows such a conditionally stable characteristic. Curves (a)
and (b) give two other cases in which the loop gain has simply been

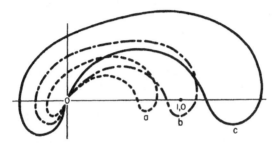

Fig.3 Conditional stability

reduced. According to Nyquist's analysis (a) and (c) should represent
stable situations while (b) is unstable. The conditionally stable situation was difficult to achieve in practice, however, because if one applied power to such an amplifier the loop had, in a sense, to grow continuously from zero. The loop must encircle the 1,0 point in the
course of this growth, resulting in self-oscillations which then usually
persisted. But Peterson, Kreer and Ware were able, by using low
frequencies and very fast switching, to demonstrate that the predic-
tions of Nyquist's criterion for the various cases illustrated in Fig. 3
were indeed exactly right.

I would like also to mention the supporting work done by my colleague, L. A. MacColl, in the same years. Most of you are no doubt
familiar with his book on servos which appeared after the war. What
I have reference to here, however, is work, much of it unpublished,
that he did in the 1920's and early 1930's on the subject of signalling,
and its relation to Fourier and Laplace transform techniques. In this
work, MacColl made much use of the systematic tools in complex
function theory, and I think that his example first led many of us to
realize that the complex variable was the natural basis for analysis of
most linear circuit problems. In particular, I think that it was MacColl's
identification of the Nyquist criterion with one form of the Cauchy integral theorem which gave us our best and broadest attack on the stability question.

Since this is a paper of reminiscences, it may be timely now to
say a few words about my own participation in the field. It came about
by accident and in some ways was never full-fledged. When I first began working on network theory, the most important devices in the
field were filters, much used to keep the channels in carrier telephone

10   ACTIVE NETWORKS AND FEEDBACK SYSTEMS

systems apart, and equalizers, which were intended to give attenuation characteristics varying inversely with those of transmission lines
so that the total characteristics would be flat. Some equalizers can
be fixed to take account of the average characteristic of the line, while
others must be variable to account for changes with temperature or
time, or local conditions in individual repeater sections.

With the burgeoning of carrier systems, the filter problem
seemed to be the most critical, and I spent a number of years trying
to improve the selectivity, or the impedance, or the phase delay, or
whatnot of contemporary filters. As projected transmission systems
became more ambitious, however, the importance of an adequate design technology for equalizers became more apparent, and beginning
in the early 1930's I spent much of my time on this subject.

The importance of the equalizer problem stems in effect from
the same tyranny of numbers that made the repeater problem more
and more serious as transmission systems became more ambitious.
As we have already seen, increasing over-all line loss meant not only
that more repeaters were needed but also that individual repeaters
had to meet higher and higher standards of quality. But by the same
token, the various jobs that equalizers have to do also tend to go up
with line loss, and each individual equalizer must be designed with
higher and higher precision if the over-all characteristic is to remain
flat within prescribed limits. Thus in Table I the 500 to 1 ratio between the over-all losses in the first and last line roughly measures
not only the severity of the repeater problem in the two cases but that
of the equalizer problem as well.

As it turns out, equalizer theory can be closely related to the
theory of active networks and feedback amplifiers because of the similarity between the general analytic conditions on such structures* and
the stability criterion. At first, however, I had no such additional
fields in mind, but merely plodded through a long program intended
to reformulate certain areas of network theory related to equalizers
as a study of the analytic behavior of some particular classes of rational functions in the complex plane. This seems pretty obvious now,
but it had not been done in any systematic manner at the time, although
MacColl and others had pointed the way. Out of the study came a sort
of algebra of the transmission characteristics of dissipative networks,
including as a special case the isolation of the minimum phase condition, which is of course essential for any unambiguous formulation of

---

*For the sake of simplicity the term "equalizer" has been used to include not
only the so-called "constant-R" circuits of conventional design practice, but also any
other structure, such as a β circuit or interstage network, which can be used to give a
varying loss characteristic under reasonably precise design control. Since the constant-
R property is chiefly useful to prevent interaction between successive units of a cas-
caded structure, a function which may also be performed, for example, by the isolating
properties of the active elements in an amplifier circuit, this sort of generalization is a
quite natural one.

FEEDBACK   11

the general relations between attenuation and phase in physical circuits. I need not review the material now, since much of it is scat-
tered here and there in my book. Perhaps the hardest job was to con-
vince myself and then my colleagues that poles and zeros in the complex plane were satisfactory design parameters. In standard filter
theory, one never deals with anything but real frequencies because
reactive circuits always have real resonances. As I recall, it took
some time to persuade several of our regular network designers that
a "resonance" at a complex frequency was anything but nonsense.

This work on equalizers led, in 1934, to my first direct encounter with the feedback problem. To begin with, I was asked to design
a variable equalizer to compensate over a considerable range for the
expected effect of temperature variations in a coaxial line. This is
not quite as simple a problem as one might think, since the desired
proportionality among the various characteristics has to be realized
on a log amplitude rather than an arithmetic amplitude scale. In due
course, however, I was able to find a solution which was at least good
enough to go on with.

Then came the real problem. It was also proposed that the
equalizer should be inserted in the feedback path of a feedback amplifier, which was otherwise already designed, without causing instability. I sweated over this problem for a long time without success. This
additional requirement on the equalizer was the straw that broke the
camel's back. At length, in desperation, I began modifying the amplifier proper rather than trying to tinker further with my equalizer.
First, I found myself tinkering up the input circuit, then the output circuit, then the two inter-stages. Finally, after I had in effect redesigned
the complete feedback loop, I found I could obtain a solution.

The technique that finally resulted was, of course, essentially
the one later described in my book. As you see, it came about as a
more or less accidental consequence of work on a particular assign-
ment, rather than as an attack, ab initio, on the general feedback
amplifier design problem. If you know the technique you will probably
also recognize that it is still the technique of an equalizer designer.
The solution is expressed in terms of a family of attenuation and phase
characteristics guaranteed to be physically realizable pairs, on the
assumption that the approximation of such characteristics is a relatively straightforward matter. I believe this is a valid assumption for
anyone who has actually had experience with equalizer or related network design problems. After all, it is what one does every day. On
the other hand, I can imagine that the situation may well seem baffling
to someone without such a background. At any rate, my book is essentially a text on equalizers and other dissipative networks, with incidental reference to amplifiers, rather than the other way around, and perhaps might have better been written with the equalizer point of view
more obviously in the foreground.

The first amplifier representing the full realization of the technique

12   ACTIVE NETWORKS AND FEEDBACK SYSTEMS

was constructed in 1936, but I did not drop out of the feedback field
until the onset of war activities, about 1940. The intervening four
years were spent in various ways. Some of this time went to detailed
refinements in design methods and applications to other feedback situations. A good deal of it went to the preparation of my book, which
was originally an informal, in-house text. However, I also spent a
good deal of effort in trying to beat the game; in other words, in trying
to find ways of achieving more feedback than the design technique as I
had so far developed it would allow. With the tubes available at that
time, the design equations seemed to promise rather meager possi-
bilities for certain broadband applications, and a way of getting more
feedback would have been quite helpful.

   Since my analysis had originally postulated single-loop, absolutely
stable systems, the natural place to look was in either multiple-loop
or conditionally stable structures. I looked at both, without much real
success. Figure 4 shows the sort of configuration that I thought of for

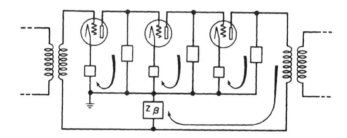

Fig.4 Multi-feedback amplifier

multi-feedback purposes. It consisted of a main feedback loop which
would break down at high frequencies into a number of local loops,
where the asymptotic high frequency transmission characteristics
could be expected to be better than they were around the main loop.
There are some interesting problems of stability associated with such
a system and I was able to make the idea work for a main loop and a
single subordinate loop. A really effective technique for the coordi-
nated design of a system containing several subordinate loops, however, was beyond me.

   The ideas I had on the exploitation of conditional stability are
illustrated in Fig. 5.[5] As I mentioned previously, conditional stability
is normally avoided because the amplifier is too likely to lock into a
sing when power is applied, or to lapse into a sing as a result of a
momentary interruption in the power supply, even if initial operation
can be achieved. It occurred to me, however, that it might be possible for the amplifier to put itself on and off the line under the control of its own sing and thus effectively detour around the critical point

FEEDBACK   13

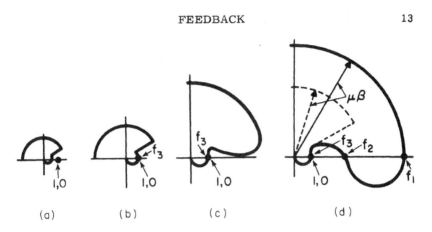

Fig.5 Stability diagram for self-adjusting amplifier

The concept is illustrated in the figure. As the $\mu\beta$ loop gain increases a sing starts at some frequency f_3 well beyond the band [see Fig.5(b)]. Through a suitable auxiliary circuit the sing current controls the shape of the $\mu\beta$ characteristic so that with further growth in the loop gain it gradually becomes conditionally stable. At the same time the auxiliary circuit limits the gain at the sing frequency so that the sing amplitude stays within tolerable bounds. Two stages in the growth of the characteristic after the initiation of sing are shown in Figs. 5(c) and 5(d).

The action required from the control circuit may seem rather elaborate, but the necessary characteristics are not really difficult to obtain when the problem is translated into requirements suitable for equalizer techniques. Figure 6 gives one solution. The two series

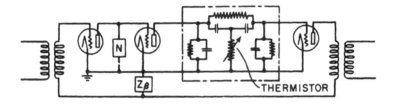

Fig.6 Control circuit for self-adjusting amplifier

condensers in the bridged-T are too small to divert much cur- rent within the operating band, but their admittances become signi- ficant at the frequency of the control sing. Thus, the sing current

14   ACTIVE NETWORKS AND FEEDBACK SYSTEMS

passes through them and the thermistor to ground. The consequent
self-heating of the thermistor reduces its value enough to limit the
sing amplitude to moderate values. At the same time, the related
phase and loss changes at other frequencies are of the sort needed to
change the original $\mu\beta$ characteristic into the desired conditionally
stable characteristic. If tube gains are reduced or if power momen-
tarily fails for any reason, the thermistor regains its normal resistance and a locked-in sing in the conditionally stable situation is
avoided.

This system actually worked and gave the expected improvement,
so far as reductions in the effects of variations in μ circuit gain were
concerned. In modulation reduction, the advantage realized was only
partial. This may well have been an accidental consequence of the
fact that the amplitude of the control sing was fairly large. Before
we had an opportunity to go exhaustively into more sensitive circuits,
however, the war emergency intervened and the experiment had to be
discontinued. The circuit now appears to be of mostly historical interest, as a sort of forerunner of modern adaptive control systems.

The idea I was looking for all this time and never found was
finally furnished by my colleague, Brockway McMillan, after the war.[6]
One embodiment is shown in Fig. 7 and an alternative in Fig. 8. An
exact analysis of either structure is complicated, but one can see approximately how the circuits work from the following: Let us suppose
for a moment that the lead to the lower amplifier in Fig. 7 is discon-
nected. Then the upper structure $(\mu_1\beta_1)$ is a normal feedback amplifier which can be analyzed by conventional means. We are concerned
with the response of the circuit to two voltages, the signal voltage E_S
applied to the input terminals, and a noise and distortion voltage, E_D,
which we can localize in the output plate circuit. It is convenient to
define E_S and E_D as the "effective" voltages which would be measured on a closed-loop basis, equal in each case to the related open-
loop value divided by the feedback factor $1 + \mu_1\beta_1$. E_S will, of
course, produce a corresponding signal voltage $\mu_1 E_S$ in the output
circuit, and E_D a related distortion voltage $\beta_1 E_D$ at the input. We
note that if $\mu_1\beta_1 \gg 1$, as would normally be the case in the useful
band, the ratio of the output to the input signal voltage is much greater
than the ratio of the output distortion voltage to the distortion voltage
at the input.

Now let the connection to the second amplifier (μ_2, β_2) be re-
stored, so that the "effective" voltages E_S and $\beta_1 E_D$ at the input of
the first amplifier are applied to the second amplifier also. If
$\mu_2\beta_2 \gg 1$, the gain of the second amplifier will be approximately
$1/\beta_2$, so that the corresponding outputs will be E_S/β_2 and $\beta_1 E_D/\beta_2$.
Clearly, if we set $\beta_1 = \beta_2$, the signal voltage output will be a negligible fraction of the output from the first amplifier, but the distortion
output will be equal to that from the first amplifier and can be made
to cancel it by a proper choice of polarities in the output combining

FEEDBACK   15

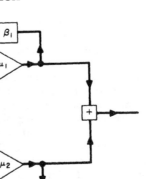

Fig.7 Negative feedback with compensation-I

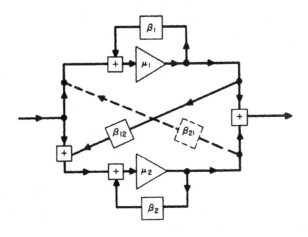

Fig.8 Negative feedback with compensation-II

circuit. The circuit of Fig. 8 can be analyzed similarly. In each case
the structures can be extended systematically by the addition of further amplifiers (μ_3, β_3) and so on, to carry the cancellation process
forward to still higher levels. In principle, the circuits provided a
solution to the problem I struggled with unsuccessfully for so long:
that of getting unlimited reduction in distortion at the cost of buying
more active elements.

   The curious feature of these circuits is the fact that they represent,

16   ACTIVE NETWORKS AND FEEDBACK SYSTEMS

in a way, a return to some of Black's early ideas. Thus the final distortion reduction depends upon a cancellation between the outputs of two identical amplifiers. The constituent amplifiers, however, now are of feedback type. This has the advantage that the critical identity between characteristics now resides in the β circuits and does not involve active elements, so that it should be easier to achieve in practice. In addition, of course, it gives us the normal feedback reduction of distortion as a starting point for the cancellation process. Perhaps the circuits can best be described as a combination of the principles involved in the early Black amplifiers of Figs. 1(a) and 1(b).

After 25 years then, the feedback art came full cycle with the invention of the McMillan amplifier and my account of "The History of an Idea" is closed.

As I remarked earlier, I will not try to conclude with a systematic critique of the field as it now looks to me. Perhaps, as a returned traveller to my old home town, however, you will permit me to give voice to one or two general impressions. I would like to begin, then, by expressing my admiration for the progress that the town has made, the fine new buildings, the fine new stores, and the general air of briskness and prosperity that permeates it. Also, however, I would like to express one or two misgivings.

The first is that the old town seems a bit crowded. Having too much company in network research was nothing that one had to worry about in the 1930's. It was a lonesome occupation, but with compensating advantages. For example, as first comers, we could choose the best building sites — that is, the high ground of mathematics and physics which promised the firmest foundations and the most sweeping outlooks — for our theoretical structures. But I am not sure that workers in the field today are quite so lucky. There can be very little mathematical terrain left over which holds the promise that function theory held for us. Moreover, even with all the new areas which have been opened up, one wonders whether there are enough really good problems to go around among all the workers now in the field, or whether effort may not be lost by duplication and micro-engraving.

My other concern has to do with the general structure of the field. Some symposia programs I have seen (not including the present one) seem to lack any clear theoretical cohesion or focus. They remind me of the way in which my home town now sprawls along the highway, or perhaps more accurately, of the classic description of Los Angeles as 27 suburbs in search of a city.

Some of this lack of cohesion is probably traceable to the original junction of the regulator field with feedback theory as it grew out of communication engineering. The two fields are really quite different. In the orthodox regulator problem, only one or a few modes of behavior are envisaged. Great stress is laid on analyzing exactly what the system may do in these circumstances, or perhaps in devising particular components which will aid it in its operation.

FEEDBACK   17

The heart of communication engineering, on the other hand, resides in the fact that we are dealing with the response of relatively complex systems to very complex ensembles of messages which are individually unpredictable but which can be dealt with in terms of some general defining characteristics. This is essentially what was meant by characterizing the early feedback structures as "wave amplifiers, " capable of responding to any signal lying within a prescribed band. We have come to understand this aspect of communication engineering, however, much better with the development of information theory since the war. As information theory takes us into more and more subtle logical characterizations of the message ensemble, we are able to deal with more and more complex situations. We seem to be drawing closer and closer to the stuff of which "intelligent" behavior is made. From this point of view, I would say that the "communication engineering" side of the modern control field would include all the topics in the present symposium but the most characteristic subjects are perhaps adaptive systems and systems with complex digital processing in the feedback loop.

Both the communication engineering and the regulator approaches are, of course, of great importance, but they are obviously quite different in fundamental intellectual texture. One feels that the war emergency produced a sort of shotgun marriage between two incompatible personalities. Shotgun marriages do sometimes work fairly well. They legitimize the offspring and they provide homes in which the children can grow up. This, however, does not necessarily mean that the principals of the marriage are ever completely in harmony. After the children have reached maturity, they have no real need to stay together. This marriage has lasted twenty years; perhaps an amicable divorce is in order.

REFERENCES

1. U.S. Patent 1,686,792.
2. See, e.g., U.S. Patent 2,102,671 or Bell System Technical Journal, January 1934.
3. Bell System Technical Journal, January 1932.
4. Bell System Technical Journal, October 1934.
5. U.S. Patent 2,367,711.
6. U.S. Patent 2,748,201.

FORCED OSCILLATIONS IN A CIRCUIT WITH NON-LINEAR RESISTANCE (RECEPTION WITH REACTIVE TRIODE)*

by Balthasar van der Pol

ONE of the most elegant applications of the geometric and analytic methods described in our foregoing essay is to the famous equation of van der Pol, $u'' + \lambda(u^2 - 1)u' + u = 0$. Although an equivalent equation had been investigated by Rayleigh, it was the study of the multivibrator, a basic electronic circuit, which brought this equation into prominence, and focussed attention once again upon the many fascinating and difficult problems connected with the study of periodic solutions of nonlinear differential equations.

Much more complex is the study of the solution of the van der Pol equation with a forcing term, $u'' + \lambda(u^2 - 1)u' + u = a \cos \omega t$. See:

J. E. Littlewood and M. Cartwright, "Forced Oscillations in Nonlinear Systems," *Contributions to the Theory of Nonlinear Oscillations*, Annals of Mathematics Studies No. 20, pp. 149–241. Princeton: Princeton University Press, 1950.

For different approaches to the study of the periodic solution of the van der Pol equation, and similar equations, when λ is small, see:

N. Krylov and N. Bogoliubov, *Introduction to Nonlinear Mechanics*, Annals of Mathematics Studies No. 11. Princeton: Princeton University Press, 1943.

N. N. Bogoliubov and Yu. A. Mitropolskii, "The Method of Integral Manifolds in Nonlinear Mechanics," *Proc. Symp. Nonlinear Osc.*, Kiev, 1961.

R. Bellman and J. M. Richardson, "Renormalization Techniques and Mean Square Averaging—I: Deterministic Equations," *Proc. Nat. Acad. Sci. USA*, Vol. 47, 1961, pp. 1191–1194.

N. Minorsky, *Nonlinear Oscillations*. Princeton. D. Van Nostrand, 1962.

Rayleigh's work is described in his classical and very readable book, *The Theory of Sound*, Vol. 1, pp. 76–81, reprinted by Dover Publications, New York, 1945.

For an introduction to the theory of periodic solutions of differential equations, see:

S. Lefschetz, *Differential Equations: Geometric Theory*. New York: Interscience Publishers, 1957.

---

* From *The London, Edinburgh, and Dublin Philosophical Magazine and Journal of Science*, Vol. 3, 1927, pp. 65–80.

Forced Oscillations in a Circuit with non-linear Resistance.
(Reception with reactive Triode.) By BALTH. VAN DER POL,
Jun., *D.Sc.**

§ 1. *Introduction.*

WHEN an E.M.F. B sin ωt acts on a circuit consisting
of a self-inductance L, capacity C, and ohmic resistance r, free and forced oscillations are generally set up, the
amplitude b of the forced oscillation being given by

$$b = \frac{\text{B}}{\sqrt{\left(\text{L}\omega - \frac{1}{\text{C}\omega}\right)^2 + r^2}}.$$

Should the system, in addition, be in resonance with the
external E.M.F. so that

$$\text{L}\omega - \frac{1}{\text{C}\omega} = 0,$$

the resultant amplitude would have the value

$$b = \frac{\text{B}}{r}.$$

It follows from this that when r approaches zero value, the
amplitude b would become infinite.

A circuit electrically coupled to a triode with reaction may
be regarded as a system the resistance of which varies with
the reaction, the resistance being positive when the reaction
coupling is loose. The circuit will commence to oscillate,
however, when the point of critical reaction is passed and
would then act as if the resistance were negative. At the
critical value of the reaction the resistance is zero. (The
term resistance here denotes what might conveniently be
called the " differential resistance " as defined by $\frac{d\text{V}}{di}$, which
expression has the dimension of a resistance analogous
to $\frac{\text{V}}{i}$).

According to the above elementary consideration, should
the critical point of the reaction be approached, even the
least external E.M.F., tuned to the frequency of the circuit,

---

* This paper was first published in the Dutch language (1924) in
Tijdschr. van het Nederlandsch Radiogenootschap, October 1924.

66   Dr. B. van der Pol *on Forced Oscillations*

would set up forced oscillations, the amplitude of which would tend to increase to infinity. This is of course contrary to actual experience.

The bends of the characteristics of the triode limit in practice the resultant amplitude. We are therefore forced to consider the resistance of the circuit as being dependent on the existing amplitude.

The deduction of this non-linear resistance from the triode characteristics has already been fully shown in previous publications * and need not be further gone into at present. If the coupling is increased past the critical reaction coupling, the resistance becomes negative and, considered linearly, the amplitude of the forced oscillation should commence to decrease again in accordance with the equation :

$$b = \frac{B}{\sqrt{(-r)^2}} = \frac{B}{r}.$$

Linearly considered, therefore, a negative resistance would not impart greater signal strength than a positive resistance of equal magnitude, and in order to obtain maximum signal strength it would apparently only be necessary to reduce the resistance to a minimum.

The free oscillations, however, which automatically fade away in the case of a positive resistance, increase exponentially when the resistance is negative. The interference of the free oscillations would become apparent for the first time at the moment the resistance becomes zero.

The experiments by Vincent, Möller, Miss Leishon, Mercier, Rossman and Zenneck †, which are fully confirmed by accurate observations made on wireless reception using a reactive triode, show, on the contrary, that a circuit *does not* oscillate spontaneously when the reaction passes the critical point of a triode system, when the latter is tuned for the reception of signals.

It can be demonstrated especially in the case of wireless telephonic reception of a carrier wave, that the reaction may be brought past the critical point without spontaneous free oscillations being set up in the circuit, which is under the influence of the carrier wave. As soon as the influence of

---

* Van der Pol, *Tijdschr. v. h. Ned. Radiogen.* i. p. 1 (1920). Van der Pol, Radio Review, i. p. 701 (1920). Appleton and Van der Pol, Phil. Mag. xlii. p. 201 (1921).
+ Vincent, Proc. Phys. Soc. Lond. xxxii. p. 84 (1920). Möller, *Jahrb. f. drahtl. Tel.* xvii. p. 269 (1921). Miss Leishon, Phil. Mag. xlvi. p. 686 (1923). Mercier, *C. R.* clxxiv. p. 448 (1922). Zenneck, *Jahrb. f. drahtl. Tel.* xxiii. p. 47 (1924).

in a Circuit with non-linear Resistance.   67

the carrier wave ceases, the circuit commences to generate free oscillations. This phenomenon may also be observed in the following simple way.

A receiving circuit will oscillate spontaneously when detuned during the reception of a carrier wave, or during telephonic reception, when the reaction coupling is greater than the critical. This is proved by the production of an audible combination tone built up by the detection (quadratic terms) from the free and forced oscillations.

If the tuning is now brought closer to resonance, the combination tone will become lower and then suddenly disappear, though the frequency still differs from zero. The combination tone is therefore not heard close to the point of resonance. This is not due to the frequency being too low to be heard, but to the actual disappearance of this beat note, caused by the *absence of free oscillations,* forced oscillations only being present in the circuit.

The explanation of this essentially non-linear phenomenon must be sought in the fact that the representative point passes at a given period through a section of the triode-characteristic where it encounters a smaller slope, so that the resistance is no longer negative but positive. This occurs with forced oscillations when the reaction coupling is greater than the critical. The effective influence is that the free oscillation is discouraged to build up, for, what might conveniently be termed " the remaining average resistance," may under these circumstances no longer be negative.

This *suppression of the free by the forced oscillations* must be ascribed to the interaction of both types of oscillation *. This interaction is entirely absent in linear circuits but is, as has already been explained elsewhere, of vital importance in the production of triode oscillations. There can be no question here of the free oscillations being " taken along by " the forced oscillation, as is sometimes expressed in articles on this subject, although in passing we may as well point out that the presence of forced oscillations renders the free oscillations subject to a frequency-correction, which behaves as if the forced " attracted " the free frequency.

Six years ago the problem of forced oscillations in a circuit with non-linear resistance was investigated for the first time by the author †. The differential equation

$$\ddot{v} + \phi(v)\dot{v} + \omega_0^2 v = \omega_1^2 B \sin \omega_1 t,$$

* The synchronous time-keeping of two clocks hung on the same wall, as long ago observed by Huygens, is another example of the suppression of free by forced oscillations.

† Van der Pol, *Tijdschr. v. h. Ned. Radiogen.* i. p. 1 (1920).

deduced at the time, again forms the basis of the present investigation. Our remarks, however, were then confined solely to dealing with the case where the resistance remained positive. At a later date and in collaboration with Dr. Appleton, the general theoretical problem where the resistance could also be negative was investigated, the results of which are to be found in an article by Dr. Appleton [*]. The present article, however, is more general and gives a more detailed account of the experiments.

In conclusion it should be pointed out that, although the conception of free and forced oscillations is essentially linear, this interpretation has been adhered to in the present non-linear investigation in view of the fact that, when the logarithmic increments or decrements are small, as is generally the case, the resistance-correction of the frequency would be negligible and the non-linear resistance characteristic would merely find expression as a first approximation in limiting the amplitudes.

§ 2. The differential equation for the anode voltage of a triode receiver acted upon by an external periodic E.M.F. is expressed by

$$\ddot{v} + \phi(v)\dot{v} + \omega_0^2 v = \omega_1^2 B \sin \omega_1 t, \quad . \quad . \quad . \quad (1)$$

in which B denotes the amplitude of the external E.M.F., due for example to a signal, $\omega_0^2 = \dfrac{1}{CL}$ the square of the angular frequency of the receiver, and ω_1 the angular frequency of the external E.M.F.

Furthermore,

$$\int \phi(v)\,dv = \frac{rv}{L} + \frac{\psi(v)}{C},$$

in which r denotes the ohmic resistance of the L, C, r circuit, and $\psi(v)$ the " oscillation characteristic " [†] of the reactive triode.

Eliminating all secondary considerations having no direct bearing on the present investigation, the problem resolves itself into the study of an electric disturbance in a circuit as indicated in fig. 1.

To simplify matters it is assumed that the triode is adjusted to work on a symmetrical portion of the oscillation characteristic, such that no even terms [‡] occur in the powers

[*] Appleton, Proc. Cambr. Phil. Soc. xxiii. p. 231 (1923).
[†] Appleton and Van der Pol, Phil. Mag. xliii. p. 179 (1921).
[‡] The quadratic term βv^2 is of no importance in the first approximation, when considering the amplitudes. The principal influence of the term βv^2 is found in detection and modulation.

in a Circuit with non-linear Resistance.   69

of the series for $\psi(v)$, and we confine ourselves to the terms :

$$\phi(v) = -\alpha + 3\gamma v^2, \quad \ldots \quad \ldots \quad (2)$$

in which α and γ are positive quantities.

The fundamental differential equation now becomes

$$\ddot{v} - \alpha\dot{v} + \gamma\frac{dv^3}{dt} + \omega_0^2 v = B\omega_1^2 \sin\omega_1 t. \quad \ldots \quad (1)$$

If the third non-linear term, $\gamma\dfrac{dv^3}{dt}$, were absent, formula (1) is reduced to the well-known differential equation for forced oscillations.

Fig. 1.

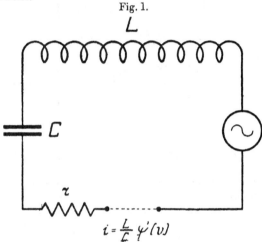

$$i = \frac{L}{C}\,\psi'(v)$$

The new term γ, which follows logically from the characteristic curves, suddenly introduces all the peculiarities discussed in § 1.

§ 3. The case wherein the receiving circuit is tuned, or nearly so, to the received oscillations is of prime interest. Hence we try for the general solution of (1) the expression

$$v = b_1 \sin\omega_1 t + b_2 \cos\omega_1 t, \quad \ldots \quad \ldots \quad (3)$$

in which b_1 and b_2 are functions of t but only slowly variable, viz. :

$$\dot{b}_1 << \omega_1 b_1,$$
$$\dot{b}_2 << \omega_1 b_2,$$
and
$$\ddot{b}_1 << \omega_1\dot{b}_1,$$
$$\ddot{b}_2 << \omega_1\dot{b}_2,$$

so that $\ddot{b}_1$ and $\ddot{b}_2$ may be neglected.

The solution allows for the possibility of the presence of free oscillations, by the assumption that both b's are functions of t. This will be explained more fully further on. In the term v^3 we only retain the fundamental frequency such that

$$v^3 = \tfrac{3}{4}(b_1^2 + b_2^2)(b_1 \sin \omega_1 t + b_2 \cos \omega_1 t). \quad . \quad . \quad (4)$$

By substituting (3) and (4) in (1) and separately equating the terms containing $\sin \omega_1 t$ and $\cos \omega_1 t$ to zero, we have :

$$\left.\begin{aligned}
2\dot{b}_1 + z b_2 - \alpha b_1 \left(1 - \frac{b^2}{a_0^2}\right) &= 0, \\
2\dot{b}_2 - z b_1 - \alpha b_2 \left(1 - \frac{b^2}{a_0^2}\right) &= B\omega_1,
\end{aligned}\right\} \quad . \quad . \quad . \quad (5)$$

in which

$$\left.\begin{aligned}
z &= \frac{\omega_0^2 - \omega_{12}}{\omega_1} \doteqdot 2(\omega_0 - \omega_1), \\
b^2 &= b_1^2 + b_2^2,
\end{aligned}\right\} \quad . \quad . \quad . \quad (6)$$

and

$$a_0^2 = \frac{\alpha}{\tfrac{3}{4}\gamma} \quad . \quad . \quad . \quad . \quad . \quad . \quad . \quad . \quad (7)$$

represents the amplitude of the free stationary oscillation.

§ 4. Before completing the solution of the non-linear case, two particular cases will be investigated first. In the first case we will investigate whither the present mode of solution, which differs from those usually adopted, will lead us in a linear case. To this end we assume in (7) that

$$\gamma = 0,$$

so that a_0^2 becomes ∞.

This has the effect of simplifying (5) to

$$\left.\begin{aligned}
2\dot{b}_1 + z b_2 - \alpha b_1 &= 0, \\
2\dot{b}_2 - z b_1 - \alpha b_2 &= B\omega_1
\end{aligned}\right\} \quad . \quad . \quad . \quad . \quad (5a)$$

the solutions of which are :

$$\left.\begin{aligned}
b_1 &= e^{\frac{a}{2}t}\left(C_1 \sin \frac{z}{2}t + C_2 \cos \frac{z}{2}t\right) - \frac{z\omega_1 B}{z^2 + \alpha^2}, \\
b_2 &= e^{\frac{\alpha}{2}t}\left(-C_1 \cos \frac{z}{2}t + C_2 \sin \frac{z}{2}t\right) - \frac{\alpha\omega_1 B}{z^2 + \sigma^2}
\end{aligned}\right\}.$$

in a Circuit with non-linear Resistance.   **71**

b_1 and b_2 are therefore of a slow periodicity. By substituting these in (3) we obtain the following equation for v :

$$v = C_3 e^{\frac{a}{2}t} \sin (\omega_0 t + \phi) + \frac{B\omega_1}{\sqrt{(\omega_0 - \omega_1)^2 + \alpha^2}}$$
$$\sin \left(\omega_1 t + \tan^{-1} \frac{2\alpha}{\omega_0 - \omega_1} \right),$$

which expression agrees with the exact linear solution of (1) if we neglect the frequency correction to which the free oscillations are subjected on account of the resistance. Hence the free oscillations are represented by the slow periodic part of the b's.

Secondly, we will consider the case in which the external E.M.F. B is equated to zero.

Expression (5) then leads to

$$b_1 = \frac{a_0 \cos \left(\frac{z}{2} t + \phi \right)}{\sqrt{1 + C e^{-at}}},$$

$$b_2 = \frac{a_0 \sin \left(\frac{z}{2} t + \phi \right)}{\sqrt{1 + C e^{-at}}};$$

so that in this case we obtain for the solution of the (only present) free oscillation in (1) :

$$v = \frac{a_0 \sin (\omega_0 t + \phi)}{\sqrt{1 + C e^{-at}}},$$

which agrees with our previous solution [*].

§ 5. Returning again to the solution of the general problem, we find that a particular solution of (5) is given by

$$\dot{b}_1 = \dot{b}_2 = 0,$$

so that

$$\left.\begin{array}{l} z b_2 - \alpha b_1 \left(1 - \dfrac{b^2}{a_0^2} \right) = 0, \\[2mm] - z b_1 - \alpha b_2 \left(1 - \dfrac{b^2}{a_0^2} \right) = B\omega_1 ; \end{array}\right\} \quad \ldots \ldots \quad (8)$$

whence

$$z^2 + \alpha^2 \left(1 - \frac{b^2}{a_0^2} \right)^2 = \frac{B^2 \omega_1^2}{b^2}. \quad \ldots \ldots \quad (9)$$

[*] *Tijdschr. v. h. Ned. Radiogen.* i. p. 21 (1920).

7 2   Dr. B. van der Pol *on Forced Oscillations*

In this instance we are dealing solely with the presence of forced oscillation, seeing that both b's in this particular solution are independent of the time. The free oscillations, which are represented by periodic parts of the b's, are here completely absent.

The circumstances under which this condition obtains are determined by a further stability investigation.

To effect this we will consider small variations Δb_1 and Δb_2 from the solutions b_1 and b_2 as obtained from (8), and determine whether or not these deviations approach 0 with the time.

From (5) and (8) follows a linear equation for both Δb's (with $D = \dfrac{d}{dt}$) :

$$\left[4D^2 - 4\alpha D\left(1 - \frac{2b^2}{a_0^2}\right) + a^2\left(1 - \frac{b^2}{a_0^2}\right)\left(1 - \frac{3b^2}{a_0^2}\right) + z^2\right]\left\{\begin{matrix}\Delta b_1 \\ \Delta b_2\end{matrix} = 0. \quad (10)$$

The solution (10), which has the form

$$x^2 + ex + f = 0,$$

only approaches zero with the time when

$$e > 0 \quad \text{and} \quad f > 0.$$

Hence the solution (8) is only stable when

$$b^2 > \tfrac{1}{2}a_0^2, \quad \ldots \ldots \quad (11)$$

and

$$a^2\left(1 - \frac{3b^2}{a_0^2}\right)\left(1 - \frac{b^2}{a_0^2}\right) + z^2 > 0. \quad \ldots \quad (12)$$

The first, (11), of these conditions required for the stable solution of (8, 9), *i.e.*, where the free oscillations are suppressed by the forced oscillations, may be put in words as follows :—

> *The development of the free oscillation is suppressed if the square of the resultant amplitude of the forced oscillation is greater than half the square of the amplitude which the free oscillation would attain in the absence of an external E.M.F.*

The second condition, (12), for the suppression of the free by the forced oscillation may also be written

$$\frac{db^2}{dB^2} > 0 \quad \ldots \ldots \ldots \quad (12\,a)$$

or

$$\frac{db^2}{dz^2} > 0 ; \quad \ldots \ldots \quad (12\,b)$$

in a Circuit with non-linear Resistance.   73

i. e. :

> *Free oscillations are also suppressed by the forced oscil-*
> *lations if in (9) of the three available values of b^2 of*
> *the forced oscillations, such a value is chosen that the*
> *amplitude increases with increasing external E.M.F.*
> *This is quite a plausible condition.*

Condition (12 b) simply lays down that *a stable solution is*
only obtained provided that an increased amplitude results from
an increased accuracy in tuning.

Fig. 2.

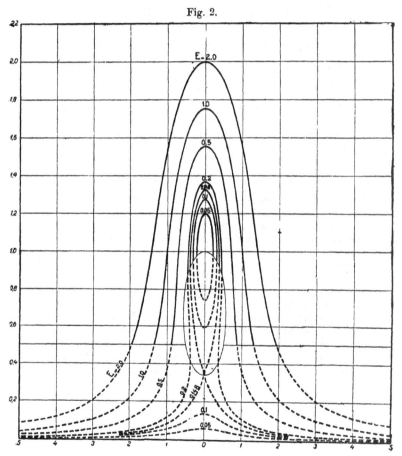

§ 6. The results obtained up to now are illustrated by a
few resonance curves given in fig. 2.

To this end (9) may be written

$$x^2 + (1-y)^2 = \frac{E}{y}, \quad \ldots \ldots \quad (13)$$

in which

$\dfrac{z}{\alpha} = x$ represents the amount of detuning,

$\dfrac{b^2}{a_0{}^2} = y$,,   the resultant amplitude,

$\dfrac{\omega_1{}^2 B}{a_0} = E$,,   the external E.M.F.

The x used in fig. 2 thus measures the degree of detuning, and the ordinate y the resultant amplitude of the forced oscillation for the following values of the external E.M.F.:

$$E = 0\cdot05 \quad 0\cdot1 \quad 0\cdot148 \quad 0\cdot2 \quad 0\cdot5 \quad 1\cdot0 \quad \text{and} \quad 2\cdot0.$$

Stable conditions are shown by full lines, whilst unstable conditions are represented by the broken lines. The unstable portions are according to (11) and (12) limited by the curves

$$y = \tfrac{1}{2} \quad \ldots \ldots \ldots \quad (14)$$

and

$$(1-3y)(1-y) + x^2 = 0, \quad \ldots \ldots \quad (15)$$

which areas are bounded by thin lines in the figure.

Fig. 2 clearly shows that for a strong signal, in which case $E=2$, the resultant amplitude attains a maximum when the circuit is in tune, dropping off to each side of the maximum in a manner practically similar to the ordinary (linear) resonance curve.

The slope of the resonance curve is, however, smaller than that obtained with a linear resistance for the same resonance amplitude.

The shape of the $E=2$, $E=1$, and $E=0\cdot5$ curves is roughly similar ; for weaker signals, *i. e.* when $E=0\cdot2$ and $0\cdot148$, the resonance curves are slightly " waisted." They are, however, only stable up to the point where the slope becomes vertical, as for stability the condition

$$\frac{db^2}{dz^2} > 0 \quad \ldots \ldots \quad (12\,b)$$

must be satisfied.

When the signals are very weak, $E=0\cdot1$ and $0\cdot05$, the resonance curve splits up into two parts, the upper part assuming an approximately elliptical form, the lower part rising to a slight maximum. This lower part, falling below

in a Circuit with non-linear Resistance.   75

the value $y=\frac{1}{2}$, is unstable as shown by (11), and need not be further discussed ; the upper portion is elliptical and only stable up to the point where the curve becomes vertical.

§ 7. We still have to consider the disturbance occurring outside the limits of stability of $b^2 =$ const. Experiments go to show that the solution is now no longer of single periodicity, and thus may be regarded as the point at which free oscillations set in. Therefore, if an unmodulated carrier wave is being received, nothing will be heard in the region for which a stable solution of $b^2 =$ const. is obtained. As soon, however, as the boundary of the stability is crossed, the familiar heterodyne note will be heard.

Fig. 2 shows that the width of this "silent region" increases with the strength of the incoming signals E. For *weak signals* this width is given by

$$x^2 + (1-y)^2 = \frac{E}{y} \quad . \quad . \quad . \quad . \quad (13)$$

and

$$x^2 + (1-y)(1-3y) = 0. \quad . \quad . \quad . \quad . \quad (15)$$

If the *signals* are *strong*, the width of the silent region is determined by

$$x^2 + (1-y)^2 = \frac{E}{y} . \quad . \quad . \quad . \quad . \quad (13)$$

and

$$y = \tfrac{1}{2}. \quad . \quad . \quad . \quad . \quad (14)$$

The last expression may also be written

$$(\omega_0 - \omega_1)^2 + \tfrac{1}{16}\alpha^2 = \frac{\omega_1^2 B}{2a_0^2}, \quad . \quad . \quad . \quad . \quad (m)$$

or, for very strong signals, where

$$\tfrac{1}{16}\alpha^2 < < (\omega_0 - \omega_1)^2, \quad . \quad . \quad . \quad . \quad . \quad (n)$$

the expression becomes

$$\frac{\omega_0 - \omega_1}{\omega_1} = \mp \frac{B}{\sqrt{2} . a_0}. \quad . \quad . \quad . \quad . \quad (16)$$

Expression (16) is therefore only valid for very strong signals, and has for this case been confirmed experimentally by Dr. Appleton [*].

* Appleton, Proc. Camb. Phil. Soc. *l. c.*

8. The disturbance beyond the limits of the silent region is given by (5), in which the b's are to be taken as functions of the time. We have not so far succeeded in obtaining a general solution for (5). It is possible, however, to determine the point at which free oscillations would set in on the borders of the silent region, by determining the small deviations Δb from b at that point. Thus for weak signals, for which (12) is valid, expression (10) gives us on the borders of the silent region :

$$\Delta b_1 = C_1 + C_2\, e^{a\left(1-\frac{2b^2}{a_0{}^2}\right)t}.$$

The second term in this expression diminishes to zero with the time, and the equilibrium at that point is therefore indifferent.

For stronger signals, the stability conditions of which are determined by (11), we obtain in the same way from equation (10) the following expression for Δb_1 at the border :

$$\Delta b_1 = C_3 \sin\left(\frac{1}{2}\sqrt{z^2-\frac{\alpha^2}{4}}\cdot t + \psi\right)$$
$$= C_3 \sin\left(\sqrt{(\omega_0-\omega_1)^2-\frac{\alpha^2}{16}}\cdot t + \psi\right).$$

The free frequency with which the circuit commences to oscillate spontaneously at the border is not ω_0, but is expressed by

$$\omega_1 - \sqrt{(\omega_0-\omega_1)^2-\frac{\alpha^2}{16}}. \quad \cdots \quad (17)$$

Hence the free frequency undergoes a correction in the direction of the forced frequency, giving the impression as if the free frequency were being attracted by the forced frequency. This correction of the free frequency has experimentally already been noted by Vincent, Möller, and Appleton.

§ 9. As a first approximation of a particular solution of (1) at a point far outside the resonance region, it is permissible to assume a linear combination of the free and forced oscillations, viz. :

$$v = a \sin(\omega_0 t + s) + b \sin(\omega_1 t + \lambda), \quad \cdots \quad (18)$$

in which a, b, s, and λ are constants.

n a *Circuit with non-linear Resistance.* 77

If only the frequencies ω_0 and ω_1 are retained in v^3, we have

$$v^3 = \tfrac{3}{4}[a(a^2 + 2b^2)\sin(\omega_0 t + s) + b(b^2 + 2a^2)\sin(\omega_1 t + \lambda)].$$
$$\cdots \quad (19)$$

On substituting (18) and (19) in (1), equating the terms containing $\sin \omega_0 t$, $\cos \omega_0 t$, $\sin \omega_1 t$, and $\cos \omega_1 t$ to zero, and neglecting the small terms, we obtain the following four equations :

$$\omega_0 - \omega_0 = 0, \qquad (a)$$

$$a\left(1 - \frac{a^2 + 2b^2}{a_0{}^2}\right) = 0, \qquad (b)$$

$$b\,\frac{z}{\omega_1} = -B\cos\lambda, \qquad (c) \qquad \cdots \quad (20)$$

$$\frac{\alpha}{\omega}b\left(1 - \frac{b^2 + 2a^2}{a_0{}^2}\right) = -B\sin\lambda. \qquad (d)$$

Equation (20 a) shows that the choice of the free oscillation frequency in (18) without correction was right as a first approximation. The symbol s does not appear in (20), which indicates that we are free to make an arbitrary choice in regard to the phase. Expression (20 b) resolves into the following two solutions :

$$a = 0, \quad \cdots \quad (21)$$

$$1 - \frac{a^2 + 2b^2}{a_0{}^2} = 0. \quad \cdots \quad (22)$$

The first solution represents the suppression of the free by the forced oscillation. This is therefore the solution applying to the silent region which has already been fully discussed; for, if (c) and (d) are squared and added together, we obtain equation (9).

Solution (22), on the other hand, admits of a finite amplitude being given to the free oscillation.

By eliminating a^2 from (22), (20 c) and (20 d) we obtain

$$z^2 + a^2\left(1 - \frac{3b^2}{a_0{}^2}\right)^2 = \frac{B^2\omega_1{}^2}{b^2}. \quad \cdots \quad (23)$$

If (23) be compared with (9) it will be seen that owing to the presence of the free oscillations the circuit behaves with regard to the forced oscillations as if γ, which indicates the change in resistance with the amplitude, had been multiplied by 3.

78   Dr. B. van der Pol *on Forced Oscillations*

Equation (22) shows moreover that the free amplitude (a^2) cannot develop to a_0^2 but only to

$$a^2 = a_0^2 - 2b^2,$$

from which it follows that, as soon as the forced amplitude equals

$$b^2 = \tfrac{1}{2}a_0^2,$$

no room remains for the free oscillation, seeing that a^2 would become negative. As soon as this occurs the free oscillation is suppressed by the forced oscillation. This agrees with the results given by equation (11) indicating the same limiting value.

§ 10. The results obtained in § 9 enable us to extend fig. 2 to the curves given in fig. 3. The mean square of the total disturbance in the presence of free oscillations is given by

$$\tfrac{1}{2}(a^2 + b^2).$$

The curves in fig. 3 are, within the silent region, the same as those shown in fig. 2 ; outside this region, where free oscillations are present, the quadratic amplitude $a^2 + b^2$ has been plotted in accordance with equations (23) and (22). Fig. 3 shows that a good connexion has been established between solutions (9) and (23), especially where the external E.M.F. is high. However, where

$$E = \frac{\omega_1^2 B^2}{a_0^2} = 0\cdot2 \text{ to } 0\cdot05$$

we encounter a small region, shown dotted in the figure, where the connexion is not complete. Only a complete solution of (5) could establish such a connexion. If fig. 3 be further compared with the experimental results obtained by Möller, it will be seen that these are in excellent agreement. (See also J. Golz, *Jahrb. Drahtl. Tel.* xix. p. 281, 1922.)

It may at first appear somewhat strange that the mean square of the disturbance, as shown in fig. 3, is sometimes smaller under the influence of an external E.M.F. than that obtained when free oscillations only are present.

In a physical way, however, this result may be roughly interpreted as follows :—If free oscillations only are present, the amplitude attained is determined by the curvature of the characteristic. If in addition a forced oscillation is present, *in a Circuit with non-linear Resistance.*   **79**

the representative point will tend to oscillate on such parts
of the characteristic where the slope is smaller, resulting in
a reduction of the mean negative resistance available for the
free oscillations. This effect more than counterbalances

Fig. 3.

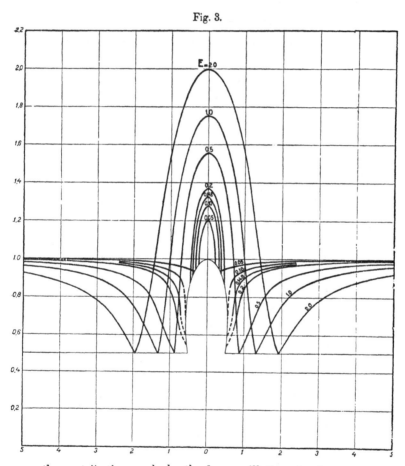

the contribution made by the free oscillations to the mean
square total amplitude, the net result of which is that the
latter decreases in consequence of the simultaneous presence
of the two oscillations.

80   *Forced Oscillations in a Circuit with non-linear Resistance.*

Summary.

If a triode system which is reaction-coupled beyond the critical point is acted upon by a signal in the form of a continuous wave, the circuit must satisfy the following equation (secondary factors being eliminated) :

$$\ddot{v} - (\alpha - 3\gamma v^2)\dot{v} + \omega_0^2 v = B\omega_1^2 \sin \omega_1 t, \quad . \quad . \quad . \quad (1)$$

in which the non-linear term containing γ is essential in order to represent the experimental facts. In the absence of an external E.M.F., *i. e.* without the right-hand term, this equation gives us a free oscillation of frequency ω_0 having a constant amplitude a_0 determined by

$$a_0^2 = \frac{a}{\frac{3}{4}\gamma}.$$

If an external E.M.F. is present, the following phenomena are observed :—

Close to the resonance region (ω_1 practically equal to ω_0) forced oscillations only are present the amplitude of which is greater than a_0. At this point the free oscillations are suppressed by the forced oscillations. Moving away from the resonance region, by varying ω_0 the amplitude of the forced oscillation decreases until it falls below a_0. Continuing this detuning, the free oscillations commence to set in with a small amplitude ; the latter are subjected, however, to a small frequency correction in the direction of the forced oscillation frequency. The mean square of the total disturbance, consisting of free and forced oscillations, is smaller than that set up by the free oscillations only.

The limits of stability of the suppression of the free oscillations are investigated in detail.

The total disturbance is generally regarded as an oscillation of forced frequency, the amplitude of which may be slowly variable.

If this amplitude is constant, the forced oscillation only is present. If in addition to the constant part of the amplitude, a slow and periodic variation of the amplitude is found, this would indicate simultaneous presence of free and forced oscillations. These oscillations react on one another due to the non-linear term containing γ in the equation, as opposed to cases of linear form in which this interaction is absent.

Physical Laboratory,
   N. V. Philips' Glowlamp Works,
      Eindhoven, Holland.

SELF-EXCITED OSCILLATIONS IN DYNAMICAL SYSTEMS POSSESSING RETARDED ACTION*

by N. Minorsky

THE ordinary differential equation may be interpreted as describing a process where interactions are instantaneous. Thus, in the equation

$$\frac{d^2x}{dt^2} = g\left(x, \frac{dx}{dt}\right), \tag{1}$$

we are tacitly assuming that the force acting at any time t depends only upon the present position and velocity. In general, the behavior of a system depends, to varying extents, upon its complete past history. Perhaps the simplest example of this is a system which is affected both by its present state and its state τ time units in the past. In this case, the equation in (1) becomes

$$\frac{d^2x}{dt^2} = g(x(t), x'(t), x(t-\tau), x'(t-\tau)). \tag{2}$$

If, as in many situations, $\tau \ll 1$, it is permissible to ignore this small time lag. In other processes, it may be essential to take it into account. This is frequently the case in the study of control and stability theory.

In this pioneer paper, Minorsky discusses the influence of time lags on the stability of a system, with particular reference to self-excited, parasitic oscillations. The novel feature of differential difference equations, such as

$$u'(t) + a_1 u(t) + a_2 u(t-1) = 0, \tag{3}$$

is that the characteristic function

$$\lambda + a_1 + a_2 e^{-\lambda} = 0 \tag{4}$$

possesses, apart from obvious trivial cases, an infinite number of roots. This fact increases the chance that a system ruled by a nonlinear differential-difference equation will possess a parasitic oscillation.

---

* From *Journal of Applied Mechanics*, Vol. 9, 1942, pp. 65–71.

The problem of determining when all the zeros of an exponential polynomial, such as

$$p(\lambda) + q(\lambda)e^{\lambda} = 0, \tag{5}$$

where $p(\lambda)$ and $q(\lambda)$ are ordinary polynomials, have negative real parts, has been treated, as mentioned previously, by Pontryagin in an important paper. For the principal results and some applications, see:

R. Bellman and K. L. Cooke, *Differential-difference Equations*. New York: Academic Press, Inc., 1962.

The stability of linear and nonlinear differential-difference equations has been, and continues to be, investigated in great detail. See the cited works of Wright, Myskis, Krasovskii, and the foregoing book, where many additional references will be found.

The study of differential-difference equations, and more generally, equations with hereditary influences, is much richer and more difficult than the study of differential equations. Relatively little is known at present and much remains to be done, particularly in connection with control theory.

Self-Excited Oscillations in Dynamical Systems Possessing Retarded Actions

By NICHOLAS MINORSKY,[1] CARDEROCK, MD.

There exists a variety of dynamical systems, possessing retarded actions, which are not entirely describable in terms of differential equations of a finite order. The differential equations of such systems are sometimes designated as hysterodifferential equations. An important particular case of such equations, encountered in practice, is when the original differential equation for unretarded quantities is a linear equation with constant coefficients and the time lags are constant. The characteristic equation, corresponding to the hysterodifferential equation for retarded quantities in such a case, has a series of subsequent high-derivative terms which generally converge. It is possible to develop a simple graphical interpretation for this equation. Such systems with retarded actions are capable of self-excitation. Self-excited oscillations of this character are generally undesirable in practice and it is to this phase of the subject that the present paper is devoted.

Introduction

THE forces, or moments, considered in dynamics as functions of parameters, which determine them, are assumed to be instantaneously in phase with these parameters.

For example, a pendulum is acted upon by a couple of gravity $C = mgl \sin \theta$, where θ is the variable parameter, i.e., the angle. The couple C is thus in phase with $\sin \theta$, or θ, for small angles. The differential equations in all such cases are either ordinary differential equations or partial differential equations, as the case may be, but of a finite order.

There exists, however, a rather restricted class of phenomena of the so-called hereditary type, in which the condition of a system at a given instant is determined not only by forces acting at that instant, but depends upon the entire history either of the preceding motion, or the preceding states of the system in general. V. Volterra has shown that such phenomena can be described mathematically in terms of "integrodifferential equations."

Another variety of systems not entirely describable in terms of differential equations of a finite order are systems possessing "retarded actions." Such systems, designated sometimes as being of a "hysteresis" type, are characterized by the fact that these retarded actions do not depend upon the entire previous history of motion, but merely reproduce variations of corresponding nonretarded actions, or forces, with a certain time lag. The differential equations of such systems are of an infinitely high order. They are designated sometimes as "hysterodifferential equations." So far no general theory for these equations has been developed.

[1] The David Taylor Model Basin, Bureau of Ships, United States Navy Department.

Presented at the National Meeting of the Applied Mechanics Division, Philadelphia, Pa., June 20–21, 1941, of THE AMERICAN SOCIETY OF MECHANICAL ENGINEERS.

Discusson of this paper should be addressed to the Secretary, A.S.M.E., 29 West 39th Street, New York, N. Y., and will be accepted until July 10, 1942, for publication at a later date. Discussion received after the closing date will be returned.

NOTE: Statements and opinions advanced in papers are to be understood as individual expressions of their authors, and not those of the Society.

An important particular case of hysterodifferential equations encountered in practical applications is when the original differential equation for unretarded quantities is a linear equation with constant coefficients and the time lags are constant. The characteristic equation, corresponding to the hysterodifferential equation in such a case, has a series of subsequent high-derivative terms which generally converge. This permits replacing the infinite series of high-derivative terms by its limit which introduces transcendental functions in its expression. Under such circumstances, it becomes possible to give a simple graphical interpretation to this equation.

The most interesting feature of such systems with retarded actions is the fact that they are capable of self-excitation with a theoretically infinite number of frequencies which are determined not only by the parameters of the dynamical system, but also by the parameter of the retarded action, i.e., its time lag.

Such self-excited oscillations are generally undesirable in practice; they are sometimes referred to as "parasitic oscillations" or "hunt."

These oscillations are investigated in this paper.

Systems With a Retarded Action

Retarded actions are generally present in any control system. In fact, a control system B is generally intended to maintain a certain state or level (e.g., of pressure, temperature, voltage, direction in space, rate of motion, etc.) of the system A to be controlled.

A departure of A from its normal state of equilibrium causes B to operate so as to eliminate this initial departure. During this "return-trip" transmission of action (from A to B) and reaction

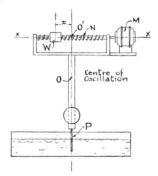

FIG. 1   SYSTEM OF PHYSICAL PENDULUM IN WHICH OCCUR SELF-EXCITED OSCILLATIONS

(from B to A), a certain time interval Δt elapses, so that the control action, arriving back at A at a certain instant t, does not correspond to the condition which A had in the past, i.e., at the instant $(t - \Delta t)$.

As an example of a system in which such self-excited oscillation was originally observed, consider a physical pendulum shown in Fig. 1. The differential equation of motion of the pendulum

A-66   JOURNAL OF APPLIED MECHANICS   JUNE, 1942

with velocity damping (e.g., produced by a paddle P moving in viscous fluid) is

$$J\ddot{\theta} + B\dot{\theta} + C\theta = 0 \dots\dots\dots\dots\dots [1']$$

where θ is the angle of oscillation and J, B, C are the moment of inertia, the coefficient of the damping, and the "effective spring constant" of the pendulum, respectively.

Assume now that, in addition to the natural damping $B\dot{\theta}$, an artificially produced damping is provided, e.g., a weight w is displaced for that purpose along the axis x-x perpendicular to the axis of the pendulum at O, by means of a motor M acting through a worm N. If the motion of the weight is so controlled[2] that its distance x from the middle point O' is proportional to angular velocity $\dot{\theta}$ of the pendulum, it is clear that the moment of the weight w applied to the pendulum is now $wx = S\dot{\theta}$, so that the differential equation of the pendulum is now

$$J\ddot{\theta} + B\dot{\theta} \pm S\dot{\theta} + C\theta = 0 \dots\dots\dots\dots [1]$$

The double sign before $S\dot{\theta}$ means that there are two ways of connecting the control to produce the moment of that kind. If the control is so connected as to increase the damping, the sign $+$ must be taken and the equation in this case becomes

$$J\ddot{\theta} + (B + S)\dot{\theta} + C\theta = 0 \dots\dots\dots\dots [2]$$

It may be seen from the form of Equation [2] that the action of the weight so controlled will merely increase the natural coefficient B of damping of the oscillating system.

However, in addition to this expected behavior, the pendulum, under certain conditions, begins to oscillate spontaneously with a much higher frequency than that corresponding to its own damped period, or to that of an externally applied disturbing moment. For example, a pendulum having a natural period of about 7 sec was found to oscillate with a period of less than 1 sec with amplitude of a fraction of a degree. This particular case will be studied in this paper as a typical case of self-excitation caused by a retarded-action phenomenon.

Hysterodifferential Equation of Motion

In the foregoing simplified treatment of the problem, it was tacitly assumed that the control system acts with an infinite rapidity, i.e., with a zero time lag. In reality, the term $S\dot{\theta}$ is a retarded quantity (in the following treatment, the bar above a symbol designates that the corresponding quantity is retarded). In the arrangement described, this means that, whereas an ideal control should introduce a lever arm x corresponding to the present instant t of motion, the real control introduces x which relates to a past instant $(t - \Delta t)$ where Δt is the time lag.

Since the angular velocity $\dot{\theta}(t)$, considered as a function of time, is a continuous function with continuous derivatives under the normal conditions (absence of impulsive forces) it can be expanded in Taylor's series

$$\bar{\dot{\theta}} = \dot{\theta} - \Delta t \ddot{\theta} + \frac{\Delta t^2}{2!}\dddot{\theta} - \frac{\Delta t^3}{3!}\theta^{\mathrm{IV}} + \dots\dots\dots\dots [3]$$

Equation [3] expresses the retarded value $\bar{\dot{\theta}}$ of angular velocity in terms of nonretarded values of $\dot{\theta}$ and its higher derivatives.

Substituting $\bar{\dot{\theta}}$ into the differential equation

$$J\ddot{\theta} + B\dot{\theta} + S\bar{\dot{\theta}} + C\theta = 0 \dots\dots\dots\dots [4]$$

---

[2] There are several ways of accomplishing this; one of the simplest consists in mounting a small constrained gyroscope on the pendulum so that it will not interfere with the motion. This gyro can control through a vacuum-tube system the speed of the motor in proportion to $\dot{\theta}$. By providing a follow-up arrangement between the excursion of the weight and the control current, the couple of the form $S\dot{\theta}$ can be produced.

Using Equation [3]

$$S\left[\dots (-1)^n \frac{\Delta t^n}{n!}\theta^{(n+1)} + \dots - \frac{\Delta t^3}{3!}\theta^{\mathrm{IV}} + \frac{\Delta t^2}{2!}\dddot{\theta} - \Delta t \cdot \ddot{\theta} \right.$$
$$\left. + \dot{\theta} \right] + J\ddot{\theta} + B\dot{\theta} + C\theta = 0 \dots [5]$$

Equation [5] is a hysterodifferential equation of motion of the pendulum. It is seen that this is an equation of an infinitely high order, with absolute values of higher-order terms decreasing with the increasing order.

It is to be noted that the new equation is still linear; there appear therefore new frequencies corresponding to an infinite number of roots of its characteristic equation.

One could proceed by limiting the number of terms of the higher order to some finite number and investigating the equation in this manner. Such procedure is generally complicated and there is no certainty as to whether the effect of terms of the higher order is negligible or not.

It is preferable for that purpose to transform this equation to a different form, owing to the fact that the series of terms of a gradually increasing order is always convergent.

Asymptotic Form of the Hysterodifferential Equation

Equation [3] can be written

$$\bar{\dot{\theta}} = \dot{\theta}\left[1 - \Delta t \frac{\ddot{\theta}}{\dot{\theta}} + \frac{\Delta t^2}{2!}\frac{\dddot{\theta}}{\dot{\theta}} - \frac{\Delta t^3}{3!}\frac{\theta^{\mathrm{IV}}}{\dot{\theta}} + \dots \right] \dots\dots [6]$$

Since the original hysterodifferential Equation [5] is linear with constant coefficients, it is satisfied by a solution $\theta = \theta e^{zt}$ where $z = (\alpha + i\omega)$; $(i = \sqrt{-1})$. Forming the subsequent derivatives entering into the ratios $\ddot{\theta}/\dot{\theta}$, $\dddot{\theta}/\dot{\theta}$, $\dots$, one finds that these ratios are z, z^2, $z^3 \dots$ so that Equation [6] becomes

$$\bar{\dot{\theta}} = \dot{\theta}\left(1 - \Delta t z + \frac{\Delta t^2}{2!}z^2 - \frac{\Delta t^3}{3!}z^3 + \dots \right) = \dot{\theta}e^{-\sigma} \dots [7]$$

where $\sigma = \Delta t \cdot z = \alpha \cdot \Delta t + i\omega \cdot \Delta t = \gamma + i\varphi$, putting $\alpha \cdot \Delta t = \gamma$ and $\omega \cdot \Delta t = \varphi$.

Substituting $\bar{\dot{\theta}} = \dot{\theta}e^{-\sigma}$ into Equation [4]

$$J\ddot{\theta} + (B + Se^{-\sigma})\dot{\theta} + C\theta = 0 \dots\dots\dots\dots [8]$$

This is another form of the hysterodifferential Equation [5] in which the infinite series of higher time derivatives arising from Taylor's expansion has been replaced by its sum, $Se^{-\sigma} \cdot \dot{\theta}$.

Substituting $\theta = \theta_0 e^{zt}$ into Equation [8]

$$[Jz^2 + (B + Se^{-\sigma})z + C]\theta_0 e^{zt} = 0 \dots\dots\dots\dots [9]$$

In order to have solutions other than trivial ones $\theta_0 = 0$, it is necessary to equate the bracket to zero, since $e^{zt} \neq 0$; this gives

$$Jz^2 + (B + Se^{-\sigma})z + C = 0 \dots\dots\dots\dots [10]$$

For self-excitation it is necessary to make $\alpha > 0$ initially. In fact, in such a case, being given a small departure θ_0 initially, the process of self-excitation will be started according to the law $\theta = \theta_0 e^{\alpha t}e^{i\omega t}$.

As regards ω, the angular frequency, it is always a positive quantity. The solution of the problem is thus reduced to that of Equation [10].

Substituting $z = (\alpha + i\omega)$

$$J(\alpha + i\omega)^2 + (B + Se^{-\gamma}e^{-i\varphi})(\alpha + i\omega) + C = 0$$

Since $e^{-i\varphi} = \cos\varphi - i\sin\varphi$, this gives rise to two equations obtained by putting separately the real and the imaginary parts equal to zero

MINORSKY—OSCILLATIONS IN DYNAMICAL SYSTEMS POSSESSING RETARDED ACTIONS   A-67

$$J(\alpha^2 - \omega^2) + B\alpha + Se^{-\gamma}(\alpha\cos\varphi + \omega\sin\varphi) + C = 0$$
$$2J\alpha\omega + B\omega + Se^{-\gamma}(\omega\cos\varphi - \alpha\sin\varphi) = 0 \quad\Big\}\,..[11]$$

Transferring terms with $Se^{-\gamma}$ on one side of Equations [11] and dividing them out (which eliminates $Se^{-\gamma}$), finally upon rearranging the terms according to the power of α

$$(J\sin\varphi)\alpha^3 + (B\sin\varphi + J\omega\cos\varphi)\alpha^2 + [(J\omega^2 + C)\sin\varphi]\alpha$$
$$+ \omega[(J\omega^2 - C)\cos\varphi + B\omega\sin\varphi] = 0 \ldots[12]$$

Equation [12] arranged according to the powers of ω is

$$(J\cos\varphi)\omega^3 + [(J\alpha + B)\sin\varphi]\omega^2 + [(J\alpha^2 - C)\cos\varphi]\omega$$
$$+ [(C + B\alpha + J\alpha^2)\alpha\sin\varphi] = 0 \ldots[13]$$

Conditions of self-excitation are fulfilled in the domain where both α and ω are real and positive.

Dividing both equations by $\cos\varphi$

$$(J\tan\varphi)\alpha^3 + (B\tan\varphi + J\omega)\alpha^2 + [(J\omega^2 + C)\tan\varphi]\alpha$$
$$+ \omega[(J\omega^2 - C) + B\omega\tan\varphi] = 0 \ldots[14]$$

$$J\omega^3 + [(J\alpha + B)\tan\varphi]\omega^2 + (J\alpha^2 - C)\omega + \alpha(C + B\alpha$$
$$+ J\alpha^2)\tan\varphi = 0 \ldots[15]$$

ZONES OF SELF-EXCITATION

Taking $\tan\varphi$ as a factor in the preceding equations

$$\tan\varphi = -\frac{\omega[J\alpha^2 + (J\omega^2 - C)]}{J\alpha^3 + B\alpha^2 + (C + J\omega^2)\alpha + B\omega^2} \ldots[16]$$

Since the observed self-oscillations are generally above the synchronism $(J\omega^2 - C) > 0$ and because $\alpha > 0$, $\tan\varphi < 0$ during the transient period of self-excitation; that is $\pi/2 < \varphi < \pi$, or $3\pi/2 < \varphi < 2\pi$.

The self-excited process eventually reaches a stable condition when $\alpha = 0$. Equation [16] shows that in this case

$$\tan\varphi_0 = -\frac{J\omega^2 - C}{B\omega} \ldots\ldots\ldots\ldots[17]$$

On the basis of Equation [16], it cannot be determined whether the self-excitation is possible below the synchronism $(J\omega^2 - C) < 0$. It will be shown in the following that only the oversynchronous case of self-excitation is possible.

During the transient period when the amplitudes are still increasing $(\alpha > 0)$, $\tan\varphi > \tan\varphi_0$; this can be seen by putting $\alpha = \infty$ in Equation [16], which gives $\tan\varphi = -0$. The self-excitation starts thus in the region of $\varphi = \pi - \epsilon$ (ϵ being a small quantity) and approaches the value of φ_0 given by Equation [17] for a steady state.

AN ALTERNATIVE METHOD

Dividing Equation [14] by $\tan\varphi$

$$J\alpha^3 + (B + J\omega\cot\varphi)\alpha^2 + (J\omega^2 + C)\alpha + \omega[(J\omega^2 - C)$$
$$\cot\varphi + B\omega] = 0 \ldots[18]$$

In this form, the first term of the equation is positive; the equation being of an odd degree in α has certainly one real root, this root is positive, as required by the condition of self-excitation if the last term is negative. Since $\omega > 0$, this condition is equivalent to

$$(J\omega^2 - C)\cot\varphi + B\omega < 0$$

whence

$$\tan\varphi > -\frac{J\omega^2 - C}{B\omega}$$

Designating the absolute value of $\tan\varphi$ by $|\tan\varphi|$ this gives

$$|\tan\varphi| < \frac{J\omega^2 - C}{B\omega}$$

For a stable state of self-excitation $\alpha = 0$; in this case the last term of Equation [18] vanishes, which gives $|\tan\varphi_0| = \dfrac{J\omega^2 - C}{B\omega}$ as previously found. Therefore, the condition for an oversynchronous excitation is

$$|\tan\varphi| \leq \frac{J\omega^2 - C}{B\omega} \ldots\ldots\ldots\ldots[19]$$

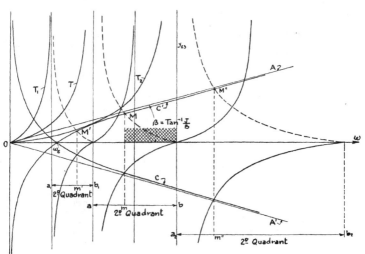

FIG. 2   GRAPHICAL REPRESENTATION OF CONDITIONS OF SELF-EXCITATION

A-68   JOURNAL OF APPLIED MECHANICS   JUNE, 1942

the sign $<$ holds for the transient condition and $=$, for the steady state.

GRAPHIC INTERPRETATION

These conditions of self-excitation are represented graphically in Fig. 2 in which $\tan \varphi = \tan(\omega \Delta t)$ is plotted against ω as curve T. $\tan \varphi$ undergoes discontinuities for $\varphi = \omega \Delta t = n\pi/2$ $(n = 1, 3, 5, \ldots)$. Absolute values of $\tan \varphi$ are also plotted in dotted lines in even quadrants where $\tan \varphi < 0$.

Straight lines OA and OA' have angular coefficients respectively equal to $+J/B$ and $-J/B$; the curves B and B' are hyperbolas $+C/B\omega$ and $-C/B\omega$ referred to the axes y and ω as asymptotes; these curves B, B' are omitted in Fig. 2.

By adding algebraically the ordinates of curves B and A' and those of curves B' and A, curves C and C' are obtained. Curve C has for its equation, $y_c = \dfrac{C}{B\omega} - \dfrac{J}{B}\omega = -\dfrac{J\omega^2 - C}{B\omega}$, and the equation for curve C' is

$$y_c' = -\frac{C}{B\omega} + \frac{J\omega}{B} = \frac{J\omega^2 - C}{B\omega}$$

Both curves intersect the abscissa axis at the point $\omega_s = \sqrt{C/J}$ which corresponds to the undamped synchronism of the system.

Condition of self-excitation of parasitic oscillations, given by the criterion, Equation [19], is found in the diagram, Fig. 2, between the ordinate y_{22} separating the second and the third quadrants and the abscissa of the point M, which is the intersection of curves C' and $|\tan \varphi|$. Within this range the inequality of Equation [19] is, in fact, satisfied.

The limit frequency is at the point m, the abscissa of M. The shaded area indicates the range of frequencies within which the self-excitation occurs in a transient state $(\alpha > 0)$.

VECTOR DIAGRAMS

The foregoing criteria show that the self-excitation above synchronism is possible in the second as well as in the fourth quadrants which means that these two oscillations may be of two different frequencies.

A further restriction to this multiplicity of possible frequencies is obtained from a consideration of conditions of dynamical equilibrium of moments applied to the dynamical system.

In view of the fact that this question resembles closely a similar question encountered in the theory of alternating currents, a parallel study of both problems is useful.

The differential equation of an electric circuit, containing an inductance L, resistance R, and capacity C, in series, and acted upon by electromotive force $E = E_0 e^{j\omega t}$ is[3]

$$L\frac{di}{dt} + Ri + \frac{1}{C}\int i\,dt = E_0 e^{j\omega t}$$

Differentiating this equation

$$L\frac{d^2 i}{dt^2} + R\frac{di}{dt} + \frac{1}{C}i = E_0 j\omega e^{j\omega t} \ldots\ldots\ldots [20]$$

The differential equation of a pendulum acted upon by an external moment $D = D_0 e^{j\omega t}$ is

$$J\frac{d^2\theta}{dt^2} + B\frac{d\theta}{dt} + C\theta = D_0 e^{j\omega t} \ldots\ldots\ldots\ldots [21]$$

It is seen that the differential Equations [20] and [21] are identical in form, with the exception that the right side of the

---

[3] In order to avoid confusion in the following equations, j designates $\sqrt{-1}$ and i the electric current in this section on "Vector Diagrams."

electrical equation figures the derivative of $E_0 e^{j\omega t}$, whereas, in the dynamical equation appears the nondifferentiated quantity $D_0 e^{j\omega t}$.

As the result of this, the real and the imaginary axes are inverted in electrical and mechanical vector diagrams. For example, the vector representing $R\,di/dt$ is along the real axis in the electrical diagram, whereas, in the mechanical diagram, the vector $B\,d\theta/dt$, representing an analogous physical quantity (the dissipation of energy), is along the imaginary axis in the mechanical diagram.

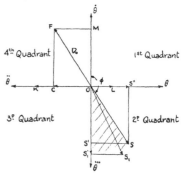

FIG. 3   DYNAMICAL-VECTOR DIAGRAM OF STEADY STATE OF SELF-EXCITATION

With this change of a formal character, the dynamical vector diagram of the steady state of self-excitation is shown in Fig. 3.

Vector $OL = C\theta_0$ is in phase with the angle θ

Vector $OM = jB\omega\theta_0$ is in phase with the angular velocity $\dot\theta$

Vector $OK = -J\omega^2\theta_0$ is in phase with the angular acceleration $\ddot\theta$

Vector $OC = OK + OL = (C - J\omega^2)\theta_0$

The vector sum $OC + CF = OC + OM = [(C - J\omega^2) + jB\omega]\theta_0$ clearly represents the left side of Equation [21] upon substituting $\theta = \theta_0 e^{j\omega t}$. Hence, OF represents the amplitude D_0 of the external moment with its phase relatively to θ.

When the self-excitation exists, the external moment D is absent. Since for dynamical equilibrium the polygon OCF is to be closed, a vector OS equal and opposite is necessary for this purpose. Thus, vector OS represents the retarded action, lagging behind the vector OM from which it is derived by an angle φ $(\pi/2 < \varphi < \pi)$. This vector OS is in the second quadrant as previously found by another method. $\tan \varphi$ is therefore negative and its absolute value is $|\tan \varphi| = OC/OM = \dfrac{J\omega^2 - C}{B\omega}$, as

found previously, Equation [19].

The vectors in the diagram represent the moments acting on the system. If they are multiplied by ω they are capable of representing the power. It is seen from the diagram Fig. 3 that the conditions of self-excitation can be formulated as follows: (1) $OM = OS'$ and (2) $OC = OS''$.

The first condition states that the power output from the system caused by the dissipation of energy $B\dot\theta$ must be equal to the power input into the system through the component $OS' = OS \cos \varphi$ of the retarded action OS. This latter component figures therefore on the same basis as a negative-resistance effect in an electric circuit; it is well known that in an electric circuit no self-excitation is possible without a negative-resistance element's being present in the circuit.

The second condition $OC = OS''$, that is, $\theta_0(J\omega^2 - C) = OS$ $\sin \varphi$ merely fixes the frequency of self-excitation.

Since OM is always directed along the imaginary axis upward, the vector OS' being equal and opposite to it, is always

MINORSKY—OSCILLATIONS IN DYNAMICAL SYSTEMS POSSESSING RETARDED ACTIONS   A-69

directed along the same axis downward. This means that the vector OS can only be either in the second or in the third quadrant. In the first case, the self-excitation is above synchronism and in the second below it.

Thus, the retarded action OS can never be either in the first or in the fourth quadrants, if the self-excitation is to occur at all. The condition of self-excitation previously found $|\tan \varphi| \leqslant \frac{J\omega^2 - C}{B\omega}$ has a simple significance in the vector diagram. In fact, since φ is contained between the limits $\frac{\pi}{2} < \varphi < \pi$ the previous condition means that the vector OS must be either somewhere inside the shaded area (during transient condition) or along the line OS. In the first case when it is, say, at OS_1 its projection is $OS_1' > OM$. In other words, in order to start the self-excitation, it is necessary to have the initial input of energy to the system greater than its dissipation. But as soon as the self-excitation is established, S_1 comes to S and $OS' = OM$, which corresponds to the steady state $\left(\tan \varphi = \frac{J\omega^2 - C}{B\omega}\right)$, as seen from the diagram; in such a case, the energy input through the retarded action $S\dot{\theta}$ is equal to its dissipation $B\dot{\theta}$

Theoretically, if the self-excitation occurs in the second quadrant, it must also occur in the sixth, tenth, etc., quadrants, by adding $2\pi n$ (n being an integer). Such higher-order self-excitations correspond however to much higher frequencies and cannot be observed in practice; the only quadrant of interest from the standpoint of self-excitation is therefore the second quadrant.

Effect of Time Lag on Self-Excitation

If the time lag is made smaller, the spacing of quadrants for $\tan \varphi$ curves becomes wider for the same frequency scale on the abscissa axis (curves T_2 in Fig. 2); for larger time lags the spacing of quadrants becomes more crowded (curves T_1).

Curves C, C', and, hence, the synchronous frequency of the system remain the same, since they do not depend upon the time lag.

Time lags, assumed in the construction of Fig. 2, are half the size for the curve T_2 and twice as large for T_1, than the time lag adopted for curve T. For this latter case, the stable frequency is at the point m. For smaller time lag (curve T_2), it is at the point m'' and for a larger lag (curve T_1) it is at m'. All three points m, m', m'' are in the second quadrant of their respective frequencies (ab, a_1b_1, and a_2b_2).

Thus, it can be seen that the frequency of self-excitation decreases with an increasing time lag and vice versa. This is generally observed in practice.

Within the frequency range a_2b_2 of the second quadrant and small time lag, the self-excited frequency m'' is nearer to the point a_2 than to the point b_2. For large time lag m' approaches the end b_1 of its frequency range a_1b_1.

Thus, with a sufficiently small time lag, the point m'' may come so close to the point a_2 that the self-excitation may disappear entirely. This will happen when the vertical projection OS' of OS (Fig. 3) will become smaller than the vector OM. At this point, the energy input into the self-excitation process becomes smaller than the losses inherent in its maintenance. It is a well-known fact that systems with very small time lags behave practically in a "dead-beat" manner.

Conversely, for a large time lag, the point m' comes closer to the point b_1 in the second quadrant (Fig. 2). In the diagram Fig. 3, the vector OS representing the retarded action approaches the downward position. In such a case the conditions for the maintenance of the hunt are the most favorable. In fact, it is

usually observed that a relatively low-frequency hunt, due to large time lags, is very persistent.

Effect of Fixed Parameters on Self-Excitation

If the time lag is small, the frequency of self-excitation is relatively high and the difference of ordinates of the curve C', Fig. 2, and its asymptote A is negligible. The angular coefficient of the asymptote A is J/B. The point of intersection M of the curve C' with the curve $|\tan \varphi|$ becomes practically identical with that of A with the same curve $|\tan \varphi|$.

The greater this ratio J/B, the nearer comes the frequency m to the critical value $\pi/2\Delta t$ at which the self-excitation ceases. The self-excited frequency thus decreases with an increasing inertia J and decreasing damping B.

Conversely, if this ratio decreases, the asymptote A rotates clockwise and the point M moves along the curve $|\tan \varphi|$ toward the point $\omega = \pi/\Delta t$ in the frequency range; the self-excited frequency increases in this case.

Thus, at a constant inertia, an increase of damping increases the frequency of the parasitic oscillation. In view of the fact that the amplitude of the parasitic oscillation is generally very small, it is difficult to provide an efficient damping.

As regards the parameter C, the spring constant of the system, it enters into the determination of the curve C' and not of its asymptote A. If, on the other hand, the time lag is small enough, the difference of ordinates between the curve C' and its asymptote A is negligible. This shows that the effect of C is also negligible, as compared to the other two factors, i.e., inertia J and damping B.

Effect of the Variable Parameter S on Self-Excitation

Parameter S, characterizing the intensity of self-excitation, that is, of the retarded action, does not appear in the preceding equations. This is due to the fact that Equations [11] were divided out so as to eliminate $Se^{-\gamma}$.

In order to be able to take into account the influence of this factor on self-excitation, Equation [10] must be studied directly. Since the determination of the regions in which the self-excitation is possible has been made already, the study can be simplified by putting $\alpha = 0$.

The condition for the existence of a steady self-excited oscillation is $\theta = \theta_0 e^{i\omega t}$, where i again designates $\sqrt{-1}$. Putting $\sigma = i\omega \Delta t = i\varphi$ in Equation [8]

$$J\ddot{\theta} + (B + Se^{-i\varphi})\dot{\theta} + C\theta = 0 \dots\dots\dots\dots [22]$$

Since $e^{-i\varphi} = \cos \varphi - i \sin \varphi$, this equation becomes

$$J\ddot{\theta} + (B + S \cos \varphi)\dot{\theta} - iS \sin \varphi \cdot \dot{\theta} + C\theta = 0 \dots [22a]$$

For a harmonic oscillation $\theta = \theta_0 e^{i\omega t}$, $\dot{\theta} = i\omega\theta$, whence $-i\dot{\theta} = \omega\theta$. This gives

$$J\ddot{\theta} + (B + S \cos \varphi)\dot{\theta} + (C + S\omega \sin \varphi)\theta = 0 \dots\dots\dots [23]$$

For a steady state of self-excitation, clearly $B + S \cos \varphi = 0$, whereas, during the transient state ($\alpha > 0$), $(B + S \cos \varphi) < 0$; thus in general $B + S \cos \varphi \leqslant 0$; that is, $\cos \varphi \leqslant -B/S$. Since both B and S are positive, φ is in the second quadrant; as to the fourth quadrant, it is ruled out as was previously explained.

Furthermore, since $\cos \varphi < 1$, $S > B$; that is, the self-excitation of oscillations cannot occur if the coefficient of intensity of the retarded action is smaller than the coefficient B of damping.

This is obvious from a physical consideration, since $S\dot{\theta}$ and $B\dot{\theta}$ are proportional to the energy input into the system and its dissipation, respectively. Clearly, the self-excitation becomes impossible if the average energy input into the system is less than its dissipation.

A-70   JOURNAL OF APPLIED MECHANICS   JUNE, 1942

For a steady state of self-excitation
$J\ddot{\theta} + (C + S\omega \sin \varphi)\theta = 0$ and, since $\sin \varphi = \sqrt{1 - (B/S)^2}$
this gives

$$J\ddot{\theta} + (C + \omega \sqrt{S^2 - B^2})\theta = 0 \dots \dots [24]$$

The steady frequency of self-excitation is thus given by the positive root of the quadratic equation

$$J\omega^2 - \omega \sqrt{S^2 - B^2} - C = 0$$

that is

$$\omega = \omega_1 + \sqrt{\omega_1{}^2 + \omega_0{}^2} \dots \dots \dots \dots [24a]$$

where $\omega_1 = \sqrt{S^2 - B^2}/2J$ and $\omega_0 = \sqrt{C/J}$.

For $S = B$, $\omega = \omega_0$; the self-excitation cannot occur below this value for S.

Thus, no self-excitation is possible below the synchronism which rules out the possibility of self-excitation in the third quadrant.

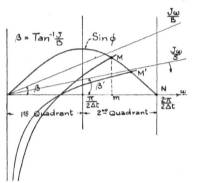

Fig. 4 Graphical Construction Showing Self-Excitation Resulting From Introduction of Sin φ

The self-excitation is thus possible only in the second quadrant ($\pi/2 < \varphi < \pi$). If S is increased, Equation [24] shows that the self-excited frequency increases as long as it remains within the range in which the self-excitation is possible, as previously investigated.

In order to be able to show how the self-excited frequency changes with S, the diagram of Fig. 2 must be changed so as to introduce this factor. Arranging the first Equation [11] according to the powers of α

$$J\alpha^2 + (B + Se^{-\gamma} \cos \varphi)\alpha + [(C - J\omega^2) + Se^{-\gamma}\omega \sin \varphi] = 0$$

The condition for $\alpha \geqslant 0$ is $[(C - J\omega^2) + Se^{-\gamma}\omega \sin \varphi] < 0$; that is

$$\sin \varphi \leqslant \frac{J\omega^2 - C}{\omega} \frac{1}{Se^{-\gamma}} \dots \dots \dots \dots [25]$$

From the second Equation [11]

$$\alpha = -\frac{\omega[B + Se^{-\gamma} \cos \varphi]}{2J\omega - Se^{-\gamma} \sin \varphi} \dots \dots \dots \dots [26]$$

Since $B + Se^{-\gamma} \cos \varphi \leqslant 0$, from Equation [26] for $\alpha > 0$, it follows that $2J\omega - Se^{-\gamma} \sin \varphi \geqslant 0$, that is

$$\sin \varphi \leqslant \frac{2J\omega^2}{\omega Se^{-\gamma}} \dots \dots \dots \dots [27]$$

Comparing inequalities of Equations [25] and [27], it is observed that the first dominates the second, because if the first is satisfied,

the second is satisfied eo ipso; hence, only the inequality, Equation [25], is of interest.

For a steady state $e^{-\gamma} = 1$ and we have

$$\sin \varphi = \frac{J\omega^2 - C}{S\omega} \text{ and } \cos \varphi = -\frac{B}{S} = -\frac{B\omega}{S\omega}$$

Thus, it is seen that, although both $\sin \varphi$ and $\cos \varphi$ depend upon S, this factor was eliminated in the expression for $\tan \varphi$, with which the previous formulas were established.

If now $\sin \varphi$ is introduced into the graphical construction of Fig. 4, instead of $\tan \varphi$, this dependence upon S will appear. This is shown in Fig. 4.

The essential difference between the two diagrams is in the fact that the angular coefficient of the asymptote is now J/S (S being a variable quantity) and not J/B as it was in the case of Fig. 2. The limit frequency is reached as soon as S reaches the value B. The point M represents this limiting frequency m. It is obtained by the intersection of curves $\sin \varphi$ and $\dfrac{J\omega^2 - C}{B\omega}$.

If S is increased, the representative point comes to M' (intersection of curves $\sin \varphi$ and $\dfrac{J\omega^2 - C}{S\omega}$, with $S > B$). With S increasing, the asymptote turns in the direction of the arrow and the stable frequency approaches the limit N at which the self-excitation disappears.

In fact, it is generally observed that, for a sufficiently intense coupling S, the parasitic oscillations disappear and the system becomes "dead beat."

Generalization

Although the preceding study relates to a rather particular case of self-excitation, a further generalization of this theory does not present any difficulties.

It must be noted that, in establishing the preceding relations, a number of simplifying hypotheses were tacitly assumed:

1 The retarded moment is a linear function of the corresponding parameter. For example, in the analysis, this function was simply $S\ddot{\theta}$, where S is a constant.

2 The dynamical system is governed by a linear differential equation with constant coefficients.

3 Time lag Δt is constant.

The first condition means that the retarded moment reproduces the retarded element of motion $(\bar{\theta}, \bar{\dot{\theta}}, \bar{\ddot{\theta}})$ without any distortion.

The second condition permits of applying the principle of superposition of component motions arising from the infinity of roots of the characteristic equation of the original Equation [5]. This, in turn, permits reducing that equation to its asymptotic form Equation [8], since the original equation is satisfied by the solution of the form $\theta_0 e^{zt}$ under such circumstances.

The last condition merely specifies the definiteness of Taylor's expansion, introduced into the problem.

These conditions introduce a certain idealization of real problems. By waiving them, the problems become more complicated, or even impossible to solve.

On the other hand, as was mentioned on several occasions, the conclusions derived from such an idealized formulation of the problem, seem to be in a fairly good agreement with the observed facts.

It is possible, under such assumptions, to extend the preceding theory to a greater variety of cases, by introducing the retarded moment $S_k\bar{\theta}^k$ into the original differential Equation [1']. The index k in this notation refers to the order of the corresponding derivative; the retarded moment previously investigated in accordance with this notation is $S_1\bar{\dot{\theta}}$, in so far as it is obtained from the first derivative.

It is possible, in this manner, to investigate the effect of other retarded couples, such as $S_0\ddot{\theta}$ (retarded angle), $S_2\ddot{\theta}$ (retarded acceleration), etc., which may appear in different problems.

The following hysterodifferential equations are written in the asymptotic form:

$$J\ddot{\theta} + B\dot{\theta} + (C + S_0 e^{-\sigma})\theta = 0 \dots \dots [I]$$

$$J\ddot{\theta} + (B + S_1 e^{-\sigma})\dot{\theta} + C\theta = 0 \dots \dots [II]$$

$$(J + S_2 e^{-\sigma})\ddot{\theta} + B\dot{\theta} + C\theta = 0 \dots \dots [III]$$

$$S_3 e^{-\sigma}\dddot{\theta} + J\ddot{\theta} + B\dot{\theta} + C\theta = 0 \dots \dots [IV]$$

Equation [II] was studied in detail in the preceding sections. Equation [I] is obtained, if the moment of the weight w in the dynamical model, shown in Fig. 1, is so timed as to be in phase with the angle θ. In Equation [III] this timing must be in phase with $\dot{\theta}$ and, in Equation [IV], in phase with $\ddot{\theta}$, and so on.

In all these cases, as well as in combinations of such cases, the procedure is the same as that previously explained in connection with Equation [II]:

1   To substitute a solution $\theta = \theta_0 e^{zt}$ which leads to the characteristic equation in asymptotic form, Equation [8].

2   To expand the characteristic equation by substituting $z = \alpha + i\omega$. In this manner, by equating separately the real and the imaginary parts to zero, two equations are obtained.

3   By eliminating $Se^{-\sigma}$ between these two equations one equation is obtained which is arranged according to the powers of α. These equations are of a gradually increasing order in terms of α, e.g., quadratic for Equation [I], cubic for Equation [II], quartic for Equation [III], etc.

4   The next step is to establish the conditions for existence of at least one positive real root for α, which determines the zones within which the self-excitation is possible.

5   The limit of self-excitation is obtained by putting $\alpha = 0$ in the equation for α.

The angle φ, obtained in this manner, always designates the lag of the retarded quantity with respect to the quantity from which it derives.

The interesting feature obtained from a comparison of different hysterodifferential Equations [I] to [IV] is that, in the vector diagram, the retarded vector $S_k\ddot{\theta}^k$ is always in the second quadrant as specified in Fig. 3. This is due to the fact that the vector diagram imposes two additional conditions for self-excitation, i.e., the energy input into the system must be equal to its dissipation and the self-excited frequency is always above synchronism.

It was shown in connection with Equation [II], in which the retarded moment is of the form $S_1\dot{\theta}$, this angle φ_1 is contained between the limits $\pi/2 < \varphi_1 < \pi$, which brings the retarded vector in the second quadrant.

On the other hand, if a similar calculation is reproduced in connection with Equation [I], it is found that $0 < \varphi_0 < \pi/2$; but, since θ lags by $\pi/2$ behind $\dot{\theta}$, the retarded vector $S_0\ddot{\theta}$ in this case is still in the second quadrant.

Thus the following important conclusions are reached:

1   The self-excitation can appear in any case, whether the retarded action is introduced through $\ddot{\theta}$, $\dot{\theta}$, or $\ddot{\theta}$, etc. But, once it is established, the retarded vector (either $S_0\ddot{\theta}$ or $S_1\dot{\theta}$, or $S_2\ddot{\theta}$) must be necessarily in the second quadrant, Fig. 3. This requires that the self-excitation must occur with such a frequency as will give $0 < \varphi < \pi/2$ in the first case ($S_0\ddot{\theta}$), or $\pi/2 < \varphi_1 < \pi$ in the second case ($S_1\dot{\theta}$), or $\pi < \varphi_2 < 3\pi/2$ in the third case ($S_2\ddot{\theta}$), etc.

In other words the relative phase angle φ between the retarded

and nonretarded quantities constitutes an additional degree of freedom of the phenomenon, by which the retarded vector $S_k\ddot{\theta}^k$ can maintain itself in the second quadrant as required by the condition of the balance of energy, by which the energy input into the system is equal to its dissipation.

2   From a formal standpoint, the hysterodifferential equations are characterized by the fact that at least one of their coefficients is complex. The reduction of this coefficient to the real quantities introduces additional terms in the coefficient of the next order (higher or lower) and this gives the conditions of self-excitation analytically.

3   There exist an infinite number of frequencies of self-excitation forming a discrete spectrum, corresponding to the infinity of roots of the characteristic equation. In practice, generally only one first frequency is observed if the time lag is small enough, the other frequencies cannot be observed on account of the smallness of their amplitudes.

CONCLUSION

Although the preceding analysis was made in connection with a pendulum of a particular type, the results obtained are quite general.

In fact, in all control systems (dynamical, electrodynamical, electrical, hydraulic, thermal, etc.), there always exists a time lag Δt in the cyclic process of their operation. A mathematical description of such a process in terms of a hysterodifferential equation simplifies the commonly established procedure of introducing "couplings" between the different parts of the system which generally lead to ordinary differential equations of a relatively high order; this usually involves considerable difficulties if one attempts to discuss the characteristic equation algebraically. Furthermore, the nature of coupling very frequently remains unknown, particularly in dynamical problems.

On the other hand, in establishing a hysterodifferential equation of a retarded control, one must know only the over-all time lag Δt, without the necessity of knowing the intermediate links in the system. Whenever this time lag is known, the application of the foregoing method is very simple.

Conversely, in a given system in which self-excited parasitic oscillations are observed, it is possible to determine the time lag from the frequency of oscillations and known parameters of the system.

It can also be shown that the self-excitation of electric oscillations (e.g., thermionic-vacuum-tube generator) can be reduced to a retarded-action phenomenon, that is, it can be expressed in terms of a hysterodifferential equation. The only difference in the case of mechanical oscillations studied is in the fact that the phase angle φ, governing the condition of self-excitation in the electrical problem, is determined by the parameters of the electric circuit (L, C, R, M) and not by the time lag Δt directly. The criteria of self-excitation are practically the same in both cases.

It is possible that the condition of self-excitation of oscillations both in electrical and mechanical systems can be better generalized on the basis of retarded-action effect common to both problems than by studying each problem, so to speak, in its own plane.

The benefit of analogies between similar problems of different engineering professions is too well known to need any additional emphasis here.

ACKNOWLEDGMENT

The writer is indebted to Prof. Sheffer, of the mathematics Department, The Pennsylvania State College, for valuable assistance during the preliminary discussion of this problem.

AN EXTENSION OF WIENER'S THEORY
OF PREDICTION*

by L. A. Zadeh and J. R. Ragazzini

ONE of the principal objectives of a scientific theory is the prediction of the future behavior of a physical system on the basis of knowledge of its past and present states. Independently, in 1940, Kolmogorov and Wiener provided new approaches to this fundamental problem; see:

A. N. Kolmogorov, "Interpolation und Extrapolation," *Bull. de l'Académie des Sciences de l'URSS*, Ser. Math., Vol. 5, 1941, pp. 3–14.

N. Wiener, *Extrapolation, Interpolation and Smoothing of Stationary Time Series.* New York: John Wiley and Sons, 1950.

An analytic formulation is as follows. Given the function $f(t)$ for $-\infty < t \leq 0$, how do we determine the function $k(t)$, so that

$$\int_0^\infty f(-s)k(s)\,ds$$

represents a good approximation to $f(h)$, for a given $h > 0$? More generally, we want to determine $k(t)$ so that the deviation

$$f(t+h) - \int_0^\infty f(t-s)k(s)\,ds \tag{1}$$

is as small as possible. It is natural to use mean-square deviation as a criterion for effectiveness of approximation. Under appropriate assumptions it turns out that $k(t)$ is determined as a solution of an integral equation, the Wiener-Hopf equation, a type of equation which first came into prominence in astrophysics.

The paper which follows represents the first significant extension of the Kolmogorov-Wiener approach, enabling many more realistic processes to be

---

* From *Journal of Applied Physics*, Vol. 21, 1950, pp. 645–665. Reprinted by permission of the American Institute of Physics.

treated. For an extension to the multidimensional case involving some difficult matrix analysis, see:

P. Masani and N. Wiener, "The Prediction Theory of Multivariate Stochastic Processes," *Acta Math.*, I, Vol. 98, 1957, pp. 111–150; II, Vol. 99, 1958, pp. 96–137.

N. Helson and D. Lowdenslager, "Prediction Theory and Fourier Series in Several Variables," *Acta Math.*, Vol. 99, 1958.

Different approaches to prediction theory using dynamic programming have been given by:

R. Bellman, *Introduction to Matrix Analysis.* New York: McGraw-Hill Book Company, Inc., 1960, pp. 154–155.

T. Odanaka, "Prediction Theory and Dynamic Programming," The *International Statistical Institute*, 32 Session, Vol. 34, 1959,

and, in great detail, with many interesting results showing the duality between control and prediction, by

R. Kalman and R. S. Bucy, "New Results in Linear Filtering and Prediction Theory," *J. Basic Engineering*, March 1961, pp. 95–108.

In particular, Kalman discusses a number of implications of the fact that information is the dual of energy.

Reprinted from JOURNAL OF APPLIED PHYSICS, Vol. 21, No. 7, 645–655, July, 1950
Copyright 1950 by the American Institute of Physics
Printed in U. S. A.

An Extension of Wiener's Theory of Prediction*

LOTFI A. ZADEH AND JOHN R. RAGAZZINI

Department of Electrical Engineering, Columbia University, New York, New York

(Received November 14, 1949)

The theory of prediction described in this paper is essentially an extension of Wiener's theory. It differs from the latter in the following respects.

1. The signal (message) component of the given time series is assumed to consist of two parts, (a) a non-random function of time which is representable as a polynomial of degree not greater than a specified number n and about which no information other than n is available; and (b) a stationary random function of time which is described statistically by a given correlation function. (In Wiener's theory, the signal may not contain a non-random part except when such a part is a known function of time.)

2. The impulsive response of the predictor or, in other words, the weighting function used in the process of prediction is required to vanish outside of a specified time interval $0 \leq t \leq T$. (In Wiener's theory T is assumed to be infinite.)

The theory developed in this paper is applicable to a broader and more practical class of problems than that covered in Wiener's theory. As in Wiener's theory, the determination of the optimum predictor reduces to the solution of an integral equation which, however, is a modified form of the Wiener-Hopf equation. A simple method of solution of the equation is developed. This method can also be applied with advantage to the solution of the particular case considered by Wiener. The use of the theory is illustrated by several examples of practical interest.

I. INTRODUCTION

PREDICTION—in the broad sense of the term—consists essentially of estimating the values of some function of time on the basis of a time series which may or may not contain random errors. For instance, a typical problem in prediction is as follows. Given a time series $e_1(t)$ which is composed of a signal $s(t)$ and a random disturbance (noise) $N(t)$; provide an estimate of $s(t+\alpha)$, α being a positive constant, as a continuous function of time. More generally, the quantity to be estimated may be a functional of $s(t)$ such as ds/dt, $\int s\,dt$, etc. In forming such estimates the mathematical operations that may be employed are usually limited by practical considerations. Thus, in most cases the operator furnishing the estimate must be linear and fixed in addition to the obvious requirement of being physically realizable. The physical counterpart of such an operator is what is commonly known as a predictor or an estimator.

It is evident that a function of time cannot be predicted intelligently unless sufficient *a priori* information is available about both the function and the errors. The nature of such information, as well as the characteristics of the signal and noise, can assume a variety of forms. Of these the more common ones have been investigated in recent years with the result that for certain classes of time series it is now possible to design

predictors which make optimum use of the *a priori* information concerning the signal and the noise. Thus, when the given time series is stationary and the correlation functions of the signal and noise are known, one can use Wiener's theory[1] to arrive at the specifications of the optimum predictor, that is, one minimizing the mean-square value of the prediction error. On the other hand, when, as is often the case in practice, the given time series is non-stationary, the available theories of prediction, notably Phillips and Weiss' theory,[2] do not lead to the best possible predictor except for a narrow class of time series. It is possible, however, to extend Wiener's theory in many different directions thereby making it applicable to a wider class of problems than is covered by either Wiener's or Phillips and Weiss' theories in their present forms. One such extension is discussed in this paper. A feature of the extension is that the signal (message) is assumed to consist of a stationary component superimposed on a non-random function of time which is known to be representable as a polynomial of degree less than or equal to a specified number n. It will also be shown that the general method developed for treating this problem can be applied with advantage to the solution of many cases of practical

---

* This work was performed in association with the Special Projects Department of The M. W. Kellogg Company for the Watson Laboratories of the Air Materiel Command.

[1] N. Wiener, "The extrapolation, interpolation, and smoothing of stationary time series," Report of the Services 19, Research Project DIC-6037, M.I.T. (February, 1942). Published in book form by John Wiley and Sons, Inc., New York (1949).

[2] R. S. Phillips and P. R. Weiss, "Theoretical calculation on best smoothing of position data for gunnery prediction," Report 532, Rad. Lab., M.I.T. (February, 1944).

interest as well as the particular case considered by Wiener.

II. FORMULATION

Consider a given time series $e_1(t)$ which is the sum of a function $s(t)$ (signal) and a stationary random disturbance $N(t)$. Let $s^*(t)$ be the quantity to be estimated and let $s^*(t)$ be related to $s(t)$ through a given linear operator $K(p)$, i.e.,

$$s^*(t) = K(p)s(t). \qquad (1)$$

$K(p)$ may be thought of as the system function of an *ideal predictor*, i.e., a predictor capable of perfect prediction in the absence of noise. In many cases, particularly those involving actual prediction, the operator $K(p)$ is not physically realizable so that the process of estimation cannot be carried out exactly even in the absence of random disturbances.

Frequently it will be convenient to use a different, though equivalent representation of Eq. (1), i.e.,

$$s^*(t) = \int_{-\infty}^{\infty} k(\tau)s(t-\tau)d\tau, \qquad (2)$$

where τ is the variable of integration and $k(t)$ represents the impulsive response of the ideal predictor. $K(p)$ shall be referred to as the ideal prediction operator. As a matter of convenience, the more common of the many possible forms which $K(p)$ and $k(t)$ can assume are given in Table I.

Like all theories of prediction, the theory to be described applies only to a special class of time series. The time series to be considered in the work which follows will be assumed to consist of a signal $s(t)$ and noise $N(t)$, with the signal being composed of a random component $M(t)$ superposed upon a non-random function of time $P(t)$, i.e.,

$$s(t) = M(t) + P(t). \qquad (3)$$

The assumptions made concerning the characteristics of $P(t)$, $M(t)$, and $N(t)$, are as follows:

(a) $P(t)$ is assumed to be representable as a polynomial in t of degree not higher than a specified number n.

(b) $M(t)$ and $N(t)$ are stationary functions of time described respectively by their auto-correlation functions $\psi_M(\tau)$ and $\psi_N(\tau)$.

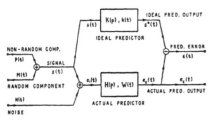

Fig. 1. Flow diagram of prediction process.

TABLE I. Common forms of the prediction operators $K(p)$ and $k(t)$.

Relation between $s^*(t)$ and $s(t)$	Significance of the quantity to be estimated	$K(p)$	$k(t)$
$s^*(t) = s(t)$	Present value of $s(t)$	1	$\delta(t)$
$s^*(t) = \dot{s}(t)$	Present value of $\dot{s}(t)$	p	$\delta^{(1)}(t)$
$s^*(t) = \ddot{s}(t)$	Present value of $\dot{s}(t)$	p^2	$\delta^{(2)}(t)$
$s^*(t) = s(t-\alpha)$	Past or future value of $s(t)$ depending respectively on whether α is a positive or negative constant	$e^{-\alpha p}$	$\delta(t-\alpha)$

Note.—$\delta(t)$ denotes a unit impulse at $t=0$, and $\delta^{(\nu)}(t)$ stands for the νth derivative of $\delta(t)$ with respect to t (time).

(c) $M(t)$ and $N(t)$ have zero mean and are uncorrelated. This assumption is introduced only for the purpose of simplification and is not essential to the analysis. The condition expressed by (c) prevails in most practical cases.

Referring to Fig. 1, these inputs are shown being applied to the actual predictor whose system function is $H(p)$ and whose impulsive response is $W(t)$. The output of the predictor, $e_2(t)$, may be expressed in operational form

$$e_2(t) = H(p)e_1(t) \qquad (4)$$

or, alternatively, in the form of a superposition integral

$$e_2(t) = \int_0^{\infty} W(\tau)e_1(t-\tau)d\tau. \qquad (5)$$

An important characteristic of the actual predictor is the so-called prediction or estimation error ϵ, which is defined as the difference between the output of the predictor and the quantity to be estimated, $s^*(t)$. Equation-wise this is:

$$\epsilon = e_2(t) - s^*(t). \qquad (6)$$

If there were no noise and if $K(p)$ were physically realizable there would be no prediction error and $H(p)$ would be identical with $K(p)$. This, of course, is the trivial case of the prediction problem. In what follows it will be assumed that either because of the presence of noise or physical unrealizability of $K(p)$, or both, $H(p)$ cannot be the same as $K(p)$.

The available *a priori* information about $s(t)$ and $N(t)$ is assumed to consist of n, $\psi_M(\tau)$, and $\psi_N(\tau)$. The problem is to specify the system function or the impulsive response of a predictor that would minimize in a certain sense the prediction error $\epsilon = e_2(t) - s^*(t)$. By analogy with Wiener's theory it will be postulated that the optimum predictor is the one in which: (a) the ensemble mean of ϵ is equal to zero (for all values of t), and (b) the ensemble variance of ϵ is a minimum. Denoting the ensemble average by the symbol $\langle \ \rangle_{Av}$, these conditions read:

(a) $\langle \epsilon \rangle_{Av} = 0$ or, equivalently, $\langle e_2(t) \rangle_{Av} \equiv \langle s^*(t) \rangle_{Av}$,   (7)

(b) $\sigma^2 = \langle \epsilon^2 \rangle_{Av} = \text{minimum}$,   (8) where σ^2, the ensemble variance of ϵ, is equal to the mean-square value of the prediction error. In what follows, conditions (a) and (b) will be used as the basis for the determination of the optimum predictor.

III. DETERMINATION OF THE IMPULSIVE RESPONSE OF THE OPTIMUM PREDICTOR

It will be recalled that the output of a predictor may be expressed in the form of a superposition integral

$$e_2(t) = \int_0^\infty W(\tau)e_1(t-\tau)d\tau \qquad (9)$$

where τ is a dummy variable and $W(t)$ represents the impulse response of the predictor. In practice it is usually found necessary to restrict the duration of sampling of the input time series to a finite constant T, meaning in other words that $W(t)$ must be zero outside the interval $0 \leq t \leq T$. To place this property in evidence Eq. (9) will be written in the following form:

$$e_2(t) = \int_0^T W(\tau)e_1(t-\tau)d\tau. \qquad (10)$$

In the limiting case where the duration of sampling is infinite $(T \to \infty)$ Eq. (10) becomes identical with Eq. (9).

By hypothesis,

$$e_1(t) = P(t) + M(t) + N(t). \qquad (11)$$

Substituting Eq. (11) into Eq. (10) and making use of the identity

$$P(t-\tau) \equiv P(t) - \tau \dot{P}(t) + \frac{\tau^2}{2!}\ddot{P}(t) + \cdots$$
$$+ (-1)^n \frac{\tau^n}{n!}P^{(n)}(t), \qquad (12)$$

it is found that $e_2(t)$ may be expressed as:

$$e_2(t) = \mu_0 P(t) - \mu_1 \dot{P}(t) + \frac{\mu_2}{2!}\ddot{P}(t) + \cdots + (-1)^n \frac{\mu_n}{n!}P^{(n)}(t)$$
$$+ \int_0^T W(\tau)M(t-\tau)d\tau + \int_0^T W(\tau)N(t-\tau)d\tau, \qquad (13)$$

where μ_0, μ_1, μ_2, etc., designate the moments of $W(t)$, i.e.,

$$\mu_\nu = \int_0^T \tau^\nu W(\tau)d\tau, \quad \nu = 0, 1, 2, \cdots n. \qquad (14)$$

Since $M(t)$ and $N(t)$ are stationary (with zero mean), it follows that the ensemble means of $e_2(t)$ and $s^*(t)$ depend only on the non-random component of the signal, i.e.,

$$\langle e_2(t)\rangle_{Av} = \int_0^T W(\tau)P(t-\tau)d\tau \qquad (15)$$

or

$$\langle e_2(t)\rangle_{Av} = \mu_0 P(t) - \mu_1 \dot{P}(t) + \frac{\mu_2}{2!}\ddot{P}(t) + \cdots$$
$$+ (-1)^n \frac{\mu_n}{n!}P^{(n)}(t), \qquad (16)$$

and

$$\langle s^*(t)\rangle_{Av} = \langle K(p)s(t)\rangle_{Av} \qquad (17)$$

or

$$\langle s^*(t)\rangle_{Av} = K(p)P(t). \qquad (18)$$

Comparing Eqs. (16) and (18), condition (a) is reduced to

$$K(p)P(t) \equiv \mu_0 P(t) - \mu_1 \dot{P}(t) + \frac{\mu_2}{2!}\ddot{P}(t) + \cdots$$
$$+ (-1)^n \frac{\mu_n}{n!}P^{(n)}(t). \qquad (19)$$

Equation (19), being an identity, determines the values of $\mu_0, \mu_1, \cdots$, and μ_n. In other words, *the ideal prediction operator $K(p)$ determines through Eq. (19) the first $n+1$ moments of the impulsive response of the optimum predictor.*

As an illustration of the foregoing statement consider a case where the quantity to be estimated is the derivative of $s(t)$, i.e., $s^*(t) = \dot{s}(t)$. For this case Eq. (19) reduces to

$$\dot{P}(t) \equiv \mu_0 P(t) - \mu_1 \dot{P}(t) + \frac{\mu_2}{2!}\ddot{P}(t) + \cdots$$
$$+ (-1)^n \frac{\mu_n}{n!}P^{(n)}(t), \qquad (20)$$

and a term by term comparison of the left-hand and right-hand sides of Eq. (20) yields:

$$\mu_0 = \int_0^T W(\tau)d\tau = 0$$

$$\mu_1 = \int_0^T \tau W(\tau)d\tau = -1$$

$$\mu_2 = \int_0^T \tau^2 W(\tau)d\tau = 0 \qquad (21)$$

$$\cdots$$

$$\mu_n = \int_0^T \tau^n W(\tau)d\tau = 0.$$

These, therefore, are the $n+1$ constraints which the impulsive response of a derivative estimating network must satisfy.

As the second example consider a case where $K(p)s(t) = s(t-\alpha)$, α being a positive or negative constant. For this case Eq. (19) reads

$$P(t-\alpha) \equiv \mu_0 P(t) - \mu_1 \dot{P}(t) + \frac{\mu_2}{2!}\ddot{P}(t) + \cdots$$

$$+ (-1)^n \frac{\mu_n}{n!} P^{(n)}(t). \quad (22)$$

Rewriting $P(t-\alpha)$ as

$$P(t-\alpha) \equiv P(t) - \alpha\dot{P}(t) + \frac{\alpha^2}{2!}\ddot{P}(t) + \cdots$$

$$+ (-1)^n \frac{\alpha^n}{n!} P^{(n)}(t), \quad (23)$$

and making in Eq. (22) a term-by-term comparison of the coefficients of $P(t)$, $\dot{P}(t)$, etc., it is easily found that:

$$\mu_0 = \int_0^T W(\tau)d\tau = 1$$

$$\mu_1 = \int_0^T \tau W(\tau)d\tau = \alpha \quad (24)$$

$$\cdots$$

$$\mu_n = \int_0^T \tau^n W(\tau)d\tau = \alpha^n,$$

which thus represent the constraints imposed upon $W(t)$ in case the quantity to be estimated is $s(t-\alpha)$.

The problem that remains to be solved is that of minimizing σ^2. For this purpose it will be necessary to develop an explicit expression for σ^2 in terms of $W(t)$ and the auto-correlation functions of the signal and noise. Assuming that condition (a) is satisfied, it follows from inspection of Eqs. (6), (13), and (19) that the prediction error is given by the expression

$$\epsilon = \int_0^T W(\tau)[M(t-\tau) + N(t-\tau)]d\tau - K(p)M(t) \quad (25)$$

or equivalently

$$\epsilon = \int_0^T W(\tau)[M(t-\tau) + N(t-\tau)]d\tau$$

$$- \int_{-\infty}^{\infty} k(\tau)M(t-\tau)d\tau \quad (26)$$

where $k(t)$ is the impulsive response of the ideal predictor. The mean-square value of ϵ may be written as

$$\sigma^2 = \langle \epsilon^2 \rangle_{Av} = \lim_{L \to \infty} \frac{1}{L} \int_0^L \epsilon^2 dt \quad (27)$$

or

$$\sigma^2 = \lim_{L \to \infty} \frac{1}{L} \int_0^L dt \left\{ \int_0^T W(\tau)[M(t-\tau) + N(t-\tau)]d\tau \right.$$

$$\left. - \int_{-\infty}^{\infty} k(\tau)M(t-\tau)d\tau \right\}^2. \quad (28)$$

A typical term of Eq. (28) such as

$$\lim_{L \to \infty} \frac{1}{L} \int_0^L dt \left[\int_0^T W(\tau)M(t-\tau)d\tau \right]^2 \quad (29)$$

is expressible in the form of a triple integral

$$\int_0^T \int_0^T d\tau_1 d\tau_2 W(\tau_1)W(\tau_2)$$

$$\times \lim_{L \to \infty} \frac{1}{L} \int_0^L M(t-\tau_1)M(t-\tau_2)dt \quad (30)$$

which in view of the definition of the auto-correlation function of $M(t)$, i.e.,

$$\psi_M(\tau) = \lim_{L \to \infty} \frac{1}{L} \int_0^L M(t)M(t-\tau)dt \quad (31)$$

may be written as

$$\int_0^T \int_0^T W(\tau_1)W(\tau_2)\psi_M(\tau_1 - \tau_2)d\tau_1 d\tau_2. \quad (32)$$

Proceeding similarly in the case of other terms, Eq. (28) reduces finally to the following expression:

$$\sigma^2 = \int_0^T \int_0^T W(\tau_1)W(\tau_2)[\psi_M(\tau_1 - \tau_2)$$

$$+ \psi_N(\tau_1 - \tau_2)]d\tau_1 d\tau_2$$

$$- 2\int_{-\infty}^{\infty} \int_0^T W(\tau_1)k(\tau_2)\psi_M(\tau_1 - \tau_2)d\tau_1 d\tau_2$$

$$+ \int_{-\infty}^{\infty} \int_{-\infty}^{\infty} k(\tau_1)k(\tau_2)\psi_M(\tau_1 - \tau_2)d\tau_1 d\tau_2 \quad (33)$$

where, to recapitulate: τ_1, τ_2 = dummy variables; $W(t)$ = impulsive response of the predictor; $\psi_M(\tau)$ = auto-correlation function of $M(t)$ [$M(t)$ is the stationary part of the input signal]; $\psi_N(\tau)$ = auto-correlation function of $N(t)$ [$N(t)$ is the input noise]; $k(t)$ = impulsive response of the ideal predictor.

Returning to the problem of minimization of σ^2 it will be noted first that the last term in Eq. (33) is independent of $W(t)$ and hence, insofar as minimization of σ^2 is concerned, need not be considered. Second, it will be recalled that $W(t)$ is subject to the $n+1$ constraints expressed by Eq. (14); therefore, the problem of minimizing σ^2 with respect to the class of $W(t)$'s satisfying Eq. (14) reduces essentially to an isoperimetric problem in the calculus of variations. Following the standard approach to such problems, one is led to minimizing the following expression:

$$I = \sigma^2 - 2\lambda_0\mu_0 - 2\lambda_1\mu_1 - \cdots - 2\lambda_n\mu_n \qquad (34)$$

or, more explicitly

$$I = \int_0^T W(\tau_1)d\tau_1 \left\{ \int_0^T W(\tau_2)[\psi_M(\tau_1-\tau_2) \right.$$

$$+ \psi_N(\tau_1-\tau_2)]d\tau_2 - 2\int_{-\infty}^{\infty} k(\tau_2)\psi_M(\tau_1-\tau_2)d\tau_2$$

$$\left. - 2\lambda_0 - 2\lambda_1\tau_1 - \cdots - 2\lambda_n\tau_1{}^n \right\} \qquad (35)$$

where the constants $\lambda_0, \lambda_1, \cdots, \lambda_n$, are the Lagrangian multipliers. Proceeding in the usual manner, that is, setting the variation of I equal to zero, it is easily found that I and hence σ^2 is a minimum provided $W(t)$ satisfies the following integral equation:

$$\int_0^T W(\tau)[\psi_M(t-\tau)+\psi_N(t-\tau)]d\tau = \lambda_0 + \lambda_1 t + \cdots$$

$$+ \lambda_n t^n + \int_{-\infty}^{\infty} k(\tau)\psi_M(t-\tau)d\tau, \quad 0 \leq t \leq T. \quad (36)$$

This equation together with the $n+1$ constraints expressed by Eq. (14) provides the basis for the determination of the optimum predictor. It will be observed that in the particular case where $n=0$, $T=\infty$, and $k(t)=\delta(t+\alpha)$ [$\delta(t)$ standing, as usual, for a unit impulse at $t=0$], Eq. (36) reduces to

$$\int_0^{\infty} W(\tau)[\psi_M(t-\tau)+\psi_N(t-\tau)]d\tau$$

$$= \psi_M(t+\alpha), \quad t \geq 0 \quad (37)$$

which is essentially the integral equation of Wiener's theory. On the other hand, in the special case where $M(t) \equiv 0$, Eq. (36) reduces to

$$\int_0^T W(\tau)\psi_N(t-\tau)d\tau = \lambda_0 + \lambda_1 t + \cdots + \lambda_n t^n,$$

$$0 \leq t \leq T \quad (38)$$

which is the integral equation of Phillips and Weiss' theory. Thus, the integral equations of Wiener's, and Phillips and Weiss' theories are special cases of Eq. (36).

IV. SOLUTION OF THE INTEGRAL EQUATION

In the general case where $\psi_M(\tau)$ and $\psi_N(\tau)$ are prescribed but otherwise arbitrary auto-correlation func-

tions, the complicated nature of the integral equation makes it appear that the solution of Eq. (36) is a formidable problem. In reality, the problem is not as difficult as it may seem, for by using a procedure to be described, the general case can be reduced to a special case which has a simple solution.

Preliminary to the discussion of this procedure it will be expedient to introduce the spectral densities of $M(t)$, $N(t)$, and $M(t)+N(t)$. Denoting these by $S_M(\omega^2)$, $S_N(\omega^2)$, and $S(\omega^2)$, respectively, and recalling that the spectral density of a function is the Fourier transform of its auto-correlation function, it follows that

$$S_M(\omega^2) = \int_{-\infty}^{\infty} \psi_M(\tau)e^{-i\omega\tau}d\tau \qquad (39)$$

$$S_N(\omega^2) = \int_{-\infty}^{\infty} \psi_N(\tau)e^{-i\omega\tau}d\tau \qquad (40)$$

and

$$S(\omega^2) = S_M(\omega^2) + S_N(\omega^2). \qquad (41)$$

Now the spectral density function $S(\omega^2)$ may be factored into the product of two conjugate factors

$$S(\omega^2) = G(j\omega) \cdot G(-j\omega) \qquad (42)$$

such that both $G(j\omega)$ and $1/G(j\omega)$ are analytic in the right half of the $j\omega$-plane. Usually $S(\omega^2)$ is assumed to be a rational function of ω^2 of the form

$$S(\omega^2) = [A(\omega^2)/B(\omega^2)], \qquad (43)$$

where $A(\omega^2)$ and $B(\omega^2)$ are polynomials in ω^2. For such cases the process of factorization is quite straightforward as can be seen from the following examples:

(a) $S(\omega^2) = \omega^2;$ $G(j\omega) = j\omega.$

(b) $S(\omega^2) = \dfrac{1}{\omega^2+\omega_0{}^2};$ $G(j\omega) = \dfrac{1}{j\omega+\omega_0}.$

(c) $S(\omega^2) = \dfrac{\omega^2+a^2}{\omega^4+b^2\omega^2+c^4};$

$$G(j\omega) = \frac{j\omega+a}{(j\omega)^2+j\omega(b^2+2c^2)^{\frac{1}{2}}+c^2}.$$

To summarize, a rational spectral density function may be written as

$$S(\omega^2) = |G(p)|^2{}_{p=j\omega} \qquad (44)$$

where $G(p)$ is of the form:

$$G(p) = \frac{Q(p)}{R(p)} = \frac{a_0+a_1p+\cdots+a_mp^m}{b_0+b_1p+\cdots+b_lp^l}, \qquad (45)$$

and the polynomials $Q(p)$ and $R(p)$ do not have any zeros in the right half of the p-plane.

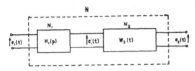

FIG. 2. Division of the predictor into the component networks N_1 and N_2.

An outline of the procedure used[3] for the solution of Eq. (36) can be best explained with reference to Fig. 2. The predictor N is assumed to be composed of two networks N_1 and N_2. The function of N_1 is to suitably modify some of the characteristics of the input time series $e_1(t)$, while that of N_2 is to provide the desired prediction through operating on the time series $e_1'(t)$, which is the output of N_1. It will be seen later that it is possible to choose N_1 in such a manner that the determination of the impulsive response of N_2 becomes an easily solvable problem. Then, once $W_2(t)$ (the impulsive response of N_2) is determined, the impulsive response of N, $W(t)$, can easily be found from the relation

$$W(t) = H_1(p)W_2(t), \qquad (46)$$

where $H_1(p)$ is the system function of N_1. The choice of $H_1(p)$ and the problem of determination of $W_2(t)$ are discussed in the sequel.

It is evident that the problem of determination of $W_2(t)$ is similar to that of the determination of $W(t)$, except that the characteristics of the input time series are different for the two problems. An inspection of the integral equation (36) shows that it can be solved rather easily when the input to the predictor consists of a polynomial in t and a stationary component whose spectral density is a polynomial in ω^2. Therefore, in order to make the determination of $W_2(t)$ a simple problem, it is necessary to provide N_2 with an input which has this property. It is not difficult to verify that such a condition will obtain if, and only if, the system function of N_1 is chosen to be

$$H_1(p) = R(p), \qquad (47)$$

where $R(p)$ is the denominator of $G(p)$ [cf. Eq. (45)]. With this choice of $H_1(p)$ the input to N_2 will consist of a polynomial in t of the same degree[4] as $P(t)$, and a stationary component $M'(t)+N'(t)$ whose spectral density is

$$S'(\omega^2) = |H_1(j\omega)|^2 S(\omega^2) \qquad (48)$$

or, in view of Eqs. (43), (44), and (47),

$$S'(\omega^2) = A(\omega^2). \qquad (49)$$

---

[3] The appendix of a report by Bode, Blackman, and Shannon, "Data smoothing and prediction in fire-control systems," Research and Development Board, Washington, D. C. (August, 1948), contains a brief exposition of a method which is similar in certain respects to the method described here.

[4] It is tacitly assumed that $R(p)$ does not have a zero at the origin or, in other words, that $S(\omega^2)$ does not have a pole at zero frequency.

where $A(\omega^2)$ is the numerator of $S(\omega^2)$. It will be noted that $A(\omega^2)$ is a polynomial of the form

$$A(\omega^2) = \gamma_0 + \gamma_1\omega^2 + \cdots + \gamma_m\omega^{2m}, \qquad (50)$$

and correspondingly the auto-correlation function of $M'(t)+N'(t)$ is

$$\psi_{M'}(\tau)+\psi_{N'}(\tau) = \gamma_0\delta(\tau) - \gamma_1\delta^{(2)}(\tau)+\cdots$$
$$+(-1)^m\gamma_m\delta^{(2m)}(\tau), \qquad (51)$$

where $\delta^{(\nu)}(\tau)$ represents the impulse function of νth order [i.e., the νth derivative of the unit impulse function $\delta(\tau)$].

In addition to $\psi_{M'}(\tau)+\psi_{N'}(\tau)$, a number of other quantities associated with the input to N_2 enter the integral equation satisfied by $W_2(t)$. The significance of each of these quantities, as well as their expressions, are as follows:

(a) $\quad S_{M'}(\omega^2) = $ spectral density of $M'(t)$
$$= S_M(\omega^2)|R(j\omega)|^2. \qquad (52)$$

(b) $\quad \psi_{M'}(\tau) = $ auto-correlation function of $M'(t)$

$$= \frac{1}{2\pi}\int_{-\infty}^{\infty} S_M(\omega^2)|R(j\omega)|^2 e^{i\omega\tau}d\omega. \qquad (53)$$

(c) $\quad k'(t) = $ ideal impulsive response for N_2
$$= [1/R(p)]k(t). \qquad (54)$$

In terms of these quantities the integral equation satisfied by $W_2(t)$ reads:

$$\int_0^\infty W_2(\tau)[\psi_{M'}(t-\tau)+\psi_{N'}(t-\tau)]d\tau$$

$$= \lambda_0' + \lambda_1't + \cdots + \lambda_n't^n$$

$$+ \int_{-\infty}^\infty k'(\tau)\psi_{M'}(t-\tau)d\tau, \quad t\geq 0. \qquad (55)$$

It will be noticed that in the case of $W_2(t)$ the upper limit of the integral is infinity, while in the case of $W(t)$ [cf. Eq. (36)] it is T. The explanation for this difference is that $W_2(t)$ need not vanish for $t>T$, even though $W(t)$ is required to do so. Thus in general $W_2(t)$ will be piecewise analytic in the interval $0<t<\infty$ as is illustrated in Fig. 3. Denoting the parts of $W_2(t)$ extending over the intervals $0\leq t\leq T$ and $T<t<\infty$ by $U(t)$ and $V(t)$, respectively, the relation connecting $W(t)$ and $W_2(t)$ [cf. Eq. (46)] may be rewritten in the following form:

$$W(t) = R(p)U(t) \qquad (56a)$$

and

$$0 = R(p)V(t). \qquad (56b)$$

These relations show that $W(t)$ is completely determined by the part of $W_2(t)$ which extends over the interval $0\leq t\leq T$; the form of $W_2(t)$ outside this interval is irrelevant to the determination of $W(t)$.

Returning to the integral equation (55), it will be noted that the range of integration $0 \leq \tau < \infty$ may be divided into two parts, $0 \leq \tau \leq T$ and $T < \tau < \infty$, involving $U(t)$ and $V(t)$, respectively. Since $V(t)$ is determined by Eq. (56b) to within a finite number of constants, the integral equation in question degenerates into an integral equation involving only $U(t)$:

$$\int_0^T U(\tau)[\psi_M'(t-\tau) + \psi_N'(t-\tau)]d\tau = \lambda_0' + \lambda_1't + \cdots$$

$$+\lambda_n't^n + \int_{-\infty}^{\infty} k'(\tau)\psi_M'(t-\tau)d\tau, \quad 0 \leq t \leq T. \quad (57)$$

Upon substitution of Eqs. (52), (53), and (54), and performing minor simplifications, Eq. (57) reads

$$\int_0^T U(\tau)[\gamma_0\delta(t-\tau) - \gamma_1\delta^{(2)}(t-\tau) + \cdots$$

$$+(-1)^m\gamma_m\delta^{(2m)}(t-\tau)]d\tau = \lambda_0' + \lambda_1't + \cdots + \lambda_n't^n$$

$$+\frac{1}{2\pi}\int_{-\infty}^{\infty} S_M(\omega^2)K(j\omega)R(-j\omega)e^{i\omega t}d\omega. \quad (58)$$

Making use of the identity

$$\int_0^T U(\tau)\delta^{(2\nu)}(t-\tau)d\tau \equiv p^{2\nu}U(t), \quad (59)$$

Equation (58) may be rewritten as

$$[\gamma_0 - \gamma_1 p^2 + \cdots + (-1)^m\gamma_m p^{2m}]U(t)$$

$$= \lambda_0' + \lambda_1't + \cdots + \lambda_n't^n$$

$$+\frac{1}{2\pi}\int_{-\infty}^{\infty} S_M(\omega^2)K(j\omega)R(-j\omega)e^{i\omega t}d\omega. \quad (60)$$

Since in this equation the left-hand side operator is simply $A(-p^2)$ [cf. Eq. (50)], the integral equation (57) finally reduces to the following differential equation:

$$A(-p^2)U(t) = \lambda_0' + \lambda_1't + \cdots + \lambda_n't^n$$

$$+\frac{1}{2\pi}\int_{-\infty}^{\infty} S_M(\omega^2)K(j\omega)R(-j\omega)e^{i\omega t}d\omega. \quad (61)$$

The general solution of this equation is of the form:

$$U(t) = A_0' + A_1't + \cdots + A_n't^n + B_1' \exp(\alpha_1 t)$$

$$+B_2' \exp(\alpha_2 t) + \cdots + B_{2m}' \exp(\alpha_{2m} t)$$

$$+\frac{1}{2\pi}\int_{-\infty}^{\infty} \frac{S_M(\omega^2)}{A(\omega^2)}K(j\omega)R(-j\omega)e^{i\omega t}d\omega,$$

$$0 \leq t \leq T, \quad (62)$$

where A_0', A_1', $\cdots$, A_n' and B_1', B_2', $\cdots$, B_{2m}' are as

yet undetermined constants, and α_1, α_2, $\cdots$, α_{2m} are the roots of the characteristic equation

$$A(-p^2) = 0. \quad (63)$$

In brief, Eq. (62) provides an explicit expression for $U(t)$ involving $2m + n + 1$ undetermined constants. Availability of such an expression reduces the problem of determination of $W(t)$ to a relatively routine matter which is discussed in the following section.

V. DERIVATION OF AN EXPLICIT EXPRESSION FOR $W(t)$

Recalling that $W(t)$ is related to $U(t)$ through the operational relation

$$W(t) = R(p)U(t), \quad (56a)$$

and substituting $U(t)$ as given by Eq. (62) into Eq. (56a), it is readily found that in the most general case $W(t)$ is given by the following expression:

$$W(t) = [u(t) - u(t-T)]\Big\{A_0 + A_1t + \cdots + A_nt^n$$

$$+B_1 \exp(\alpha_1 t) + \cdots + B_{2m} \exp(\alpha_{2m} t)$$

$$+\frac{1}{2\pi}R(p)\int_{-\infty}^{\infty} \frac{S_M(\omega^2)}{A(\omega^2)}K(j\omega)R(-j\omega)e^{i\omega t}d\omega\Big\}$$

$$+C_1\delta(t) + \cdots + C_{l-m}\delta^{(l-m-1)}(t)$$

$$+D_1\delta(t-T) + \cdots + D_{l-m}\delta^{(l-m-1)}(t-T) \quad (64)$$

where the A's, B's, C's, and D's are as yet undetermined constants, and the unit step functions $u(t)$ and $u(t-T)$ are used simply to indicate that $W(t)$ is zero outside the interval $0 \leq t \leq T$. The impulse functions contained in the expression for $W(t)$ arise from operation by $R(p)$ on the discontinuities of $U(t)$ at $t=0$ and $t=T$. It will be observed that the order of these impulse functions does not exceed $l-m-1$, which is one-half the order of the zero of $S(\omega^2)$ at infinity minus one. This is due to the fact that the first $m-1$ derivatives of $U(t)$ vanish at $t=0$ and $t=T$. It is not difficult to verify that if this would not have been the case, the mean-square error at the output of N_2 would be infinite.

Having obtained the general expression for $W(t)$ in the form of Eq. (64), there remains the problem of determination of the $2l+n+1$ unknown constants. These can be found in the following manner:

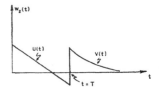

FIG. 3. Form of the impulsive response of N_2.

1. Substituting $W(t)$ as given by Eq. (64) into the integral equation (36) and requiring that the equation be satisfied identically, leads to $2l$ linear homogeneous equations in the A's, B's, C's, and D's.

2. Substituting $W(t)$ as given by Eq. (64) into the $n+1$ moment equations

$$\int_0^T \tau^\nu W(\tau)d\tau = \mu_\nu, \quad \nu = 0, 1, 2, \cdots, n \quad (65)$$

yields other $n+1$ linear equations. These $n+1$ equations, together with the $2l$ equations obtained in (1), provide a system of $2l+n+1$ linear equations in the unknown constants. Solution of this system gives the values of the A's, B's, C's, and D's and thus completes the process of determination of $W(t)$.

It should be remarked that in some cases it is advantageous to deal with the system function $H(p)$ of the predictor, rather than with its impulsive response $W(t)$. In such cases one can use a transformed form of the integral equation (36) which is as follows:

$$\frac{1}{2\pi}\int_{-\infty}^{\infty} H(j\omega)S(\omega^2)e^{j\omega t}d\omega = \lambda_0 + \lambda_1 t + \cdots + \lambda_n t^n$$

$$+\frac{1}{2\pi}\int_{-\infty}^{\infty} S_M(\omega^2)K(j\omega)R(-j\omega)e^{j\omega t}d\omega. \quad (66)$$

Using Eq. (64), the solution of this equation may be written directly as

$$H(p) = \int_0^T (A_0 + A_1 t + \cdots + A_n t^n)e^{-pt}dt$$

$$+\frac{B_1}{p+\alpha_1} + \frac{B_2}{p+\alpha_2} + \cdots + \frac{B_{2m}}{p+\alpha_{2m}}$$

$$-\left\{\frac{B_1 \exp(\alpha_1 T)}{p+\alpha_1} + \frac{B_2 \exp(\alpha_2 T)}{p+\alpha_2} + \cdots\right.$$

$$\left. +\frac{B_{2m} \exp(\alpha_{2m}T)}{p+\alpha_{2m}}\right\}e^{-pT} + \frac{1}{2\pi}R(p)\int_0^T dt e^{-pt}$$

$$\times \int_{-\infty}^{\infty} \frac{S_M(\omega^2)}{A(\omega^2)}K(j\omega)R(-j\omega)e^{j\omega t}d\omega$$

$$+C_1 + C_2 p + \cdots + C_{l-m}p^{l-m-1}$$

$$+(D_1 + \cdots + D_{l-m}p^{l-m-1})e^{-pT}. \quad (67)$$

The undetermined constants involved in this expression are found in the same manner as in the case of $W(t)$, that is, $H(p)$ as given by Eq. (67) is substituted into the integral equation (66) and the resulting expression is treated as an identity. The $2l$ linear relations between

$A_0, A_1, A_2, \cdots$, etc., which are obtained in this manner are adjoined to the $n+1$ relations resulting from Eq. (65); then the system of linear equations in the unknown constants is solved for $A_0, A_1, A_2, \cdots$, etc.

In order to facilitate application of the techniques described in the preceding sections, a summary of the procedure for the determination of $W(t)$ (or $H(p)$) is given in Section VI. Furthermore, actual use of the procedure is illustrated by a few practical examples at the end of the section.

VI. SUMMARY OF THE PROCEDURE FOR DETERMINATION OF $W(t)$ AND $H(p)$

The complete expressions for $W(t)$ (the impulsive response of the optimum predictor) and $H(p)$ (the system function of the optimum predictor) are given by Eqs. (64) and (67). In order to avoid the necessity for reference to preceding sections, the meaning of all symbols appearing in these equations is given:

$u(t)$ = unit step function.
T = duration of sampling (settling time).
$A_0, A_1, \cdots, A_n, B_1, B_2, \cdots, B_{2m}, C_1, C_2, \cdots, C_{l-m}, D_1, D_2, \cdots, D_{l-m}$ = undetermined constants.
n = degree of the polynomial component of the input signal.
$S_M(\omega^2)$ = spectral density of $M(t)$ [$M(t)$ is the stationary part of the input signal].
$S(\omega^2)$ = spectral density of $M(t)+N(t)$ [$N(t)$ is the input noise].
$A(\omega^2)$ = numerator of $S(\omega^2)$.
$B(\omega^2)$ = denominator of $S(\omega^2)$.
$Q(j\omega)$ = a factor of $A(\omega^2)$ containing all the zeros in right half of the $j\omega$-plane.
$R(j\omega)$ = a factor of $B(\omega^2)$ containing all the zeros in the left half of the $j\omega$-plane.
$2l$ = degree of $B(\omega^2)$.
$2m$ = degree of $A(\omega^2)$.
$\alpha_1, \alpha_2, \cdots, \alpha_{2m}$ = roots of the characteristic equation $A(-p^2)=0$.
$\delta(t)$ = unit impulse function.
$\delta^{(\nu)}(t)$ = νth derivative of $\delta(t)$.

The undetermined constants occurring in the expression for $W(t)$ [and $H(p)$] can be found in the following manner.

1. $W(t)$ as given by Eq. (64) is substituted into the integral equation (36) and the resulting expression is treated as an identity. This gives $2l$ homogeneous linear equations in the unknown constants. Same equations can be obtained by substituting $H(p)$, as given by Eq. (67), into the integral equation (66).

2. $W(t)$ as given by Eq. (64) is substituted into the $n+1$ constraint equations

$$\int_0^T \tau^\nu W(\tau)d\tau = \mu_\nu, \quad \nu = 0, 1, \cdots, n. \quad (14)$$

where the μ_ν are determined by the choice of the prediction operator $K(p)$ [cf. Eq. (19)]. The resulting $n+1$ linear equations in the unknown constants are adjoined to the $2l$ equations obtained from (1). The set of $2l+n+1$ linear equations thus obtained is solved for the undetermined constants $A_0, A_1, \cdots, D_{l-m}$. This concludes the process of determining $W(t)$ [or $H(p)$].

VII. ILLUSTRATIVE EXAMPLES

Example 1. Wiener's Theory

Wiener's theory is, in the main, a study of the particular case in which $P(t) \equiv 0$, $T = \infty$ and $K(p) = e^{-\alpha p}$. For this case Eq. (64) gives

$$W(t) = u(t) \left\{ \frac{1}{2\pi} R(p) \right.$$

$$\times \int_{-\infty}^{\infty} \frac{S_M(\omega^2)}{A(\omega^2)} K(j\omega) R(-j\omega) e^{j\omega t} d\omega$$

$$\left. + B_1 \exp(\alpha_1 t) + \cdots + B_{2m} \exp(\alpha_{2m} t) \right\}. \quad (68)$$

The exponential terms appearing in Eq. (68) may be made to vanish through a slight rearrangement of the factors in the first term of Eq. (68). The resulting expression for $W(t)$ is the same as that obtained by using Wiener's theory, namely,

$$W(t) = u(t) \frac{1}{2\pi} \frac{R(p)}{Q(p)}$$

$$\times \int_{-\infty}^{\infty} \frac{S_M(\omega^2)}{Q(-j\omega)} R(-j\omega) e^{j\omega(t-\alpha)} d\omega. \quad (69)$$

The rearrangement amounts, essentially, to choosing a particular solution of Eq. (61) which differs from the one chosen before by the exponential terms of Eq. (68). The same result may be achieved directly by choosing $H_1(p)$ [cf. Eq. (47)] as

$$H_1(p) = R(p)/Q(p).$$

With this choice of $H_1(p)$ [in place of the one expressed by Eq. (47)] the various quantities entering Eq. (55) become:

$$S'(\omega^2) = 1,$$
$$\psi_{M'}(\tau) + \psi_{N'}(\tau) = \delta(\tau),$$
$$S_{M'}(\omega^2) = S_M(\omega^2) |R(j\omega)/Q(j\omega)|^2,$$
$$K'(j\omega) = K(j\omega) Q(j\omega)/R(j\omega),$$

and hence the integral equation (55) reduces to

$$W_2(t) = u(t) \frac{1}{2\pi} \int_{-\infty}^{\infty} \frac{S_M(\omega^2)}{Q(-j\omega)} K(j\omega) R(-j\omega) e^{j\omega t} d\omega; \quad (55a)$$

Eq. (69) then follows immediately from the relation connecting $W(t)$ and $W_2(t)$ [cf. Eq. (46)].

Example 2

The assumptions made here are as follows:

1. $M(t) \equiv 0.$

2. $n = 1.$

3. $\psi_N(\tau) = e^{-a|\tau|}; \quad S_N(\omega^2) = \dfrac{2a}{\omega^2 + a^2}.$

The choice of the prediction operator is left open.

Solution

For this case $A(-p^2) = 2a$, and hence $\alpha_1 = \cdots = \alpha_{2m} = 0$. Also, $l = 1$, $M \equiv 0$; hence Eq. (64) gives

$$W(t) = A_0 + A_1 t + C_1 \delta(t) + D_1 \delta(t - T), \quad (70)$$

and

$$H(p) = \frac{A_0}{p}(1 - e^{-pT})$$

$$+ \frac{A_1}{p}\left(\frac{1}{p} - \frac{e^{-pT}}{p} - Te^{-pT}\right) + C_1 + D_1 e^{-pT}. \quad (71)$$

Substituting $H(p)$ as given by Eq. (71) into the integral equation

$$\frac{1}{2\pi j} \int_{-j\infty}^{j\infty} \frac{2a}{a^2 - p^2} H(p) e^{pt} dp = \lambda_0 + \lambda_1 t, \quad 0 \le t \le T, \quad (72)$$

and requiring that this equation be satisfied identically, leads to the following relations:

$$aA_0 - A_1 - a^2 C_1 = 0, \quad (73)$$

$$aA_0 + (aT+1)A_1 - a^2 D_1 = 0. \quad (74)$$

Furthermore, substituting $W(t)$ as given by Eq. (70) into the constraint equations

$$\int_0^T W(\tau) d\tau = \mu_0 \quad (75)$$

and

$$\int_0^T \tau W(\tau) d\tau = \mu_1, \quad (76)$$

yields

$$A_0 T + A_1 (T^2/2) + C_1 + D_1 = \mu_0, \quad (77)$$

and

$$A_0(T^2/2) + A_1(T^3/3) + D_1 T = \mu_1. \quad (78)$$

The unknown constants A_0, A_1, C_1, and D_1 can be readily found from the solution of Eqs. (73), (74), (77), and (78). Thus,

$$A_0 = \mu_0 \frac{4a(a^2T^2 + 3aT + 3)}{(a^2T^2 + 6aT + 12)(aT+2)}$$

$$- \mu_1 \frac{6a^2}{a^2T^2 + 6aT + 12}, \quad (79)$$

$$A_1 = -\mu_0 \frac{6a^2}{a^2T^2 + 6aT + 12} + \mu_1 \frac{12a^2}{T(a^2T^2 + 6aT + 12)}, \quad (80)$$

$$C_1 = \mu_0 \frac{2(2a^2T^2 + 9aT + 12)}{(a^2T^2 + 6aT + 12)(aT+2)}$$

$$- \mu_1 \frac{6(aT+2)}{T(a^2T^2 + 6aT + 12)}, \quad (81)$$

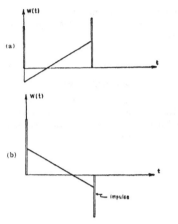

FIG. 4. Shapes of optimum weighting functions. (a) For the best value of $s(t)$. (b) For the best value of $s(t-\alpha)$.

and

$$D_1 = -\mu_0 \frac{2aT(aT+3)}{(a^2T^2+6aT+12)(aT+2)}$$

$$+\mu_1 \frac{6(aT+2)}{T(a^2T^2+6aT+12)}. \quad (82)$$

Whenever the stationary part of the input signal is zero [i.e., $M(t) \equiv 0$], the mean-square value of the prediction error assumes the simple form

$$\sigma^2 = \mu_0 \lambda_0 + \mu_1 \lambda_1 + \cdots + \mu_n \lambda_n, \quad (83)$$

which may be readily established by substituting Eq. (36) into the general expression for σ^2 [cf. Eq. (33)]. For the particular case under consideration Eq. (83) gives

$$\sigma^2 = \mu_0^2 \frac{8(a^2T^2+3aT+3)}{(a^2T^2+6aT+12)(aT+2)}$$

$$+\mu_1^2 \frac{24a}{T(a^2T^2+6aT+12)}$$

$$-\mu_0\mu_1 \frac{24a}{(a^2T^2+6aT+12)}. \quad (84)$$

The expressions given above are valid for any choice of the prediction operator. For the particular case in which the predictor is called upon to furnish the best possible estimate of the present value of $\dot{s}(t)$, the values of μ_0 and μ_1 are, respectively,

$$\mu_0 = 0$$

$$[\text{cf. Eq. (21)}] \quad (85)$$

$$\mu_1 = -1.$$

On the other hand, in the case of the estimation of $s(t+\alpha)$ (i.e., the value of $s(t)$ α-seconds in the future)

$$\mu_0 = 1$$

$$[\text{cf. Eq. (24)}] \quad (86)$$

$$\mu_1 = -\alpha.$$

The shapes of $W(t)$ for these two particular cases are illustrated in Fig. 4.

Example 3

The case to be considered here is the same as that treated in Example 2, except that the auto-correlation function of $N(t)$ is assumed to be of the form

$$\psi_N(\tau) = e^{-a|\tau|} \cos\omega_0\tau, \quad (87)$$

with the associated spectral density function being

$$S_N(\omega^2) = \frac{2a(a^2+\omega_0^2+\omega^2)}{\omega^4+2(a^2-\omega_0^2)\omega^2+(a^2+\omega_0^2)^2}. \quad (88)$$

This form of spectral density function is of considerable practical importance since it provides a reasonably good approximation to many of the actual spectra encountered in practice.

Solution

By Eq. (64), the weighting function for this case is of the form

$$W(t) = A_0 + A_1 t + B_1 e^{bt} + B_2 e^{-bt}$$

$$+C_1\delta(t) + D_1\delta(t-T), \quad 0 \leq t \leq T, \quad (89)$$

where $b = (a^2+\omega_0^2)^{\frac{1}{2}}$. Substituting $W(t)$ as given by Eq. (89) into Eq. (36), and requiring that Eq. (36) be satisfied by $W(t)$ establishes four linear algebraic equations between the six constants A_0, A_1, B_1, B_2, C_1, D_1. These are:

$$-2ab^2A_0 + 2(a^2-\omega_0^2)A_1 - b^3B_1 + b^3B_2 + 2b^4C_1 = 0,$$

$$2\omega_0^2b^2A_0 - 4a\omega_0^2A_1 + b^3(b-a)B_1 + b^3(b+a)B_2 = 0,$$

$$-2ab^2A_0 + 2(\omega_0^2-a^2-aTb^2)A_1$$

$$+b^3e^{bT}B_1 - b^3e^{-bT}B_2 + 2b^4D_1 = 0, \quad (90)$$

$$2\omega_0^2b^2A_0 + 2\omega_0^2(2a+b^2T)A_1 + b^3(b+a)e^{bT}B_1$$

$$+b^3(b-a)e^{-bT}B_2 = 0.$$

The constraints imposed by the prediction operator $K(p)$ are given by Eq. (14); they are:

$$\int_0^T W(\tau)d\tau = \mu_0, \quad (75)$$

and

$$\int_0^T \tau W(\tau)d\tau = \mu_1. \quad (76)$$

The requirement that $W(t)$ must satisfy Eqs. (75) and (76) leads to two additional linear equations. These are:

$$2bTA_0 + bT^2A_1 + 2(e^{bT} - 1)B_1 - 2(e^{-bT} - 1)B_2 \\ + 2bC_1 + 2bD_1 = 2b\mu_0,$$

and

$$3b^2T^2A_0 + 2b^2T^3A_1 + 6[e^{bT}(bT - 1) + 1]B_1 \\ - 6[e^{-bT}(bT + 1) - 1]B_2 + 6b^2TD_1 = 6b^2\mu_1. \quad (91)$$

The coefficients A_0, A_1, B_1, B_2, C_1, D_1 of Eq. (89) are the solutions of the six linear equations (90) and (91).

The expression for the mean-square value of the pre- diction error can be conveniently expressed in terms of A_0 and A_1. Thus, making use of Eq. (83) it is readily found that

$$\sigma^2 = (2a/b^2)(\mu_0A_0 + \mu_1A_1). \quad (92)$$

This completes essentially the solution of the problem.

ACKNOWLEDGMENTS

The authors wish to thank Mr. H. T. Marcy of The M. W. Kellogg Company and Mr. H. Sherman of the Watson Laboratories, AMC, for their encouragement and support of this work.

TIME OPTIMAL CONTROL SYSTEMS*

by J. P. LaSalle

ONE of the fundamental problems of control theory is that of transforming a system from one state to another in minimum time, or at minimum cost of resources in general. There are many variants of this fundamental problem and many powerful approaches now available. In classical form, the question has been treated at great length by means of the calculus of variations, and it can be handled quite directly by means of dynamic programming.

A nonclassical version was formulated by Bushaw in the following terms. Consider the linear system

$$\frac{dx}{dt} = Ax + y, \qquad x(0) = c, \tag{1}$$

where the control vector $y(t)$ is to be chosen, subject to constraints on the components of the form

$$|y_i| \leq m_i, \qquad i = 1, 2, \ldots, N, \tag{2}$$

so as to minimize the time required to drive x to the origin. It was shown by Bellman, Glicksberg, and Gross—see:

R. Bellman, I. Glicksberg, and O. Gross, "On the 'Bang-bang' Control Problem," *Q. Appl. Math.*, Vol. 14, 1956, pp. 11–18—

that certain aspects of this problem could be handled by a combination of classical analysis and modern functional analysis. The question of effective computational solution, however, is still open. Work along these lines was carried out by a number of mathematicians, notably Gamkrelidze and LaSalle. In the present paper, LaSalle greatly extends and generalizes the original results.

For an extension of some of these results to the nonlinear case, see:

E. B. Lee and L. Markus, "Optimal Control for Nonlinear Processes," *Arch. Rat. Mech. and Anal.*, Vol. 8, 1961, pp. 36–58.

---

* From *Proceedings of the National Academy of Sciences*, Vol. 45, 1959, pp. 573–577.

It is interesting to note that it was this problem which stimulated Pontryagin to develop his maximum principle.

Finally, let us note that a problem of quite similar nature, that of finding the shortest route through a network, can be handled by means of dynamic programming:

R. Bellman, "On a Routing Problem," *Q. Appl. Math.*, Vol. 16, 1958, pp. 87–90.

R. Bellman and R. Kalaba, "On k-th Best Policies," *J. Soc. Indust. and Appl. Math.*, Vol. 8, 1960, pp. 582–588,

and by several other techniques, as well. See the review papers:

R. Kalaba, "On Some Communication Network Problems," *Combinatorial Analysis*, ed. by R. Bellman and M. Hall. Providence, Rhode Island: Amer. Math. Soc., 1960.

M. Pollack, "Solutions of the k-th Best Route Through a Network—a Review," *J. Math. Anal. and Appl.*, Vol. 3, 1961, pp. 547–559.

For a general discussion of linear and nonlinear control processes from different points of view, see:

R. Bellman, I. Glicksberg, and O. Gross, *Some Aspects of the Mathematical Theory of Control Processes*, The RAND Corporation, Report R-313, 1958. (Russian translation, Moscow: 1962.)

Reprinted from the Proceedings of the NATIONAL ACADEMY OF SCIENCES
Vol. 45, No. 4, pp. 573–577.   April, 1959.

TIME OPTIMAL CONTROL SYSTEMS*

By J. P. LaSalle

RIAS, BALTIMORE, MARYLAND

Communicated by S. Lefschetz, February 27, 1959

1. *Introduction.*—It has been an intuitive assumption for some time that if a control system is being operated from a limited source of power then the system can be moved from one state to another in the shortest time by at all times utilizing properly all available power.   This hypothesis is called the "bang-bang principle." Bushaw accepted this hypothesis and in 1952 showed for some simple systems with one degree of freedom that of all bang-bang systems (that is, systems which at all times utilize maximum power) there is one that is optimal.[1]   In 1953 I made the observation that the best of all bang-bang systems, if it exists, is then the best of all systems operating from the same power source.[2]   More recently fairly general results have been obtained by Bellman, Glicksberg, and Gross[3] and later (but seemingly independently) by Krasovskii[4] and Gamkrelidze.[5]   At the 1958 International Congress of Mathematicians in Edinburgh, L. S. Pontryagin announced a "maximum principle" which is the beginning of an even more general theory.

We confine ourselves here to the time optimal problem for control systems which are linear in the sense that the elements being controlled are linear and as a function of time the control enters linearly.   The differential equation for such systems is

$$\dot{x}(t) = A(t)x(t) + B(t)u(t) + f(t), \tag{1}$$

where x and f are n-dimensional vector functions (x(t) is the state of the system at time t), A is an (n × n) matrix function, and B is an (n × r) matrix function.   Thus (1) represents the system of differential equations

$$\dot{x}_i(t) = \sum_{j=1}^{n} a_{ij}(t)x_j(t) + \sum_{k=1}^{r} b_{ik}(t)u_k(t) + f_i(t), \ i = 1, \ldots, n.$$

574   *MATHEMATICS: J. P. LaSALLE*   Proc. N. A. S.

Our ability to control the system lies in the freedom we have to choose the "steering" function u. We assume that the admissible steering functions are piecewise continuous (or measurable) and have components less than 1 in absolute value ($|u_i(t)| \leq 1$). Given an initial state x_0 and a moving particle $z(t)$, the problem of time optimal control is to hit the particle in minimum time. Let $x(t, u)$ be the solution of (1) satisfying $x(0) = x_0$. An admissible steering function u^* is *optimal* if for some $t^* > 0$, $x(t^*, u^*) = z(t^*)$ and if $x(t, u) \neq z(t)$ for $0 < t < t^*$ and all admissible u.

Bellman, Glicksberg, and Gross[3] considered the system

$$\dot{x}(t) = Ax(t) + Bu(t) \qquad (2)$$

and restricted themselves to the problem of starting at x_0 and reaching the origin in minimum time. The $(n \times n)$ matrix A is constant and its characteristic roots were assumed to have negative real parts. B was assumed to be a constant nonsingular $(n \times n)$ matrix. For some of the simplest examples of control systems the matrix B is singular, and this restriction on B is much too severe. They prove the existence of an optimal steering function, and the form for an optimal steering function is given in the proof. However, the form given for an optimal steering function does not, in general, imply that there is a bang-bang optimal steering function. Gamkrelidze[5] considered the same problem, removed the restriction that B be nonsingular, and showed for systems which are later in this paper called "normal" the existence and uniqueness of an optimal steering function. The form of the optimal steering is the same as that given by Bellman, Glicksberg, and Gross, and in this case one can conclude that the optimal steering is bang-bang. Krasovskii[4] studied the more general control system (1) and the more general control problem of hitting a moving particle. Using results of Krein on the L-problem in abstract spaces, he proved the existence of an optimal steering function for systems which we call "proper" control systems. If Krein's results are to be used without modification, the restriction to proper control systems seems to be necessary. Krasovskii states also that the optimal steering function is unique and simple examples show this to be false. Thus to date the most general bang-bang principle has been proved by Gamkrelidze.

For the more general control system (1) we show essentially that anything that can be done by an admissible steering function can also be accomplished by using bang-bang steering. This extends our result[2] and at the same time establishes the bang-bang principle for all control systems where the controlled elements are linear. This does not mean that all optimal steering functions are bang-bang. For some systems the objective can be reached in minimal time using a steering function which, during part of the time, has some zero components. We state a number of results for proper and for normal control systems which show the significance of these classifications. As in the special problem considered by Gamkrelidze, the more general normal systems have unique optimal steering functions, and in this case we have a true bang-bang principle: the only way to reach the objective in minimum time is to use the maximum available power all of the time. In Theorem 5 we give a result which should be of importance in the synthesis problem, which is the problem of determining the optimal steering u^* as a function of the state of the system. This result shows that for some systems optimal steering can be determined

Vol. 45, 1959   *MATHEMATICS: J. P. LaSALLE*   575

by what amounts to running the system backwards. This idea gives, for instance, a much simplified solution of the example solved in the paper of reference 3.

2. The General Problem.—The problem described in the introduction for the system (1) of hitting a moving particle in minimum time will be called the *general problem*. For the control system (1) the state x(t, u) of the system at time t is given by

$$x(t, u) = X(t)x_0 + X(t)\int_0^t Y(\tau)u(\tau)d\tau + X(t)\int_0^t X^{-1}(\tau)f(\tau)d\tau. \qquad (3)$$

X(t) is the principal matrix solution of $X(t) = A(t)X(t)$, and $Y(\tau) = X^{-1}(\tau)B(\tau)$. We want at some time t to have x(t) = z(t); i.e., to have

$$w(t) = \int_0^t Y(\tau)u(\tau)d\tau, \qquad (4)$$

where $w(t) = X^{-1}(t)z(t) - x_0 - \int_0^t X^{-1}(\tau)f(\tau)d\tau$. We assume throughout that A(t), B(t), and f(t) are continuous for $0 \le t < \infty$. The following Lemma states that anything that can be done by an admissible steering function can also be done by a bang-bang function. The set of admissible steering functions is the set Ω and the set of bang-bang steering functions is the set Ω^0. The set K(t) is related to the set of all states that can be reached in time t by an admissible steering function. $K^0(t)$ is similarly related to the set of states that can be reached in time t by bang-bang steering functions.[6]

LEMMA 1. *Let Ω be the set of all r-dimensional vector functions measurable on [0, t] with $|u_i(\tau)| \le 1$. Let Ω^0 be the subset of functions in Ω with $|u_i(\tau)| = 1$. Let $Y(\tau)$ be any $(n \times r)$ matrix function in $L^2([0, t])$. Define*

$$K(t) = \left\{ \int_0^t Y(\tau)u(\tau)d\tau; \qquad u \in \Omega \right\}$$

and

$$K^0(t) = \left\{ \int_0^t Y(\tau)u^0(\tau)d\tau; \qquad u^0 \in \Omega^0 \right\}.$$

Then $K^0(t)$ is closed, and $K^0(t) = K(t)$.

As a direct consequence of the Lemma we obtain an extension of the result in the paper of reference and a general bang-bang principle.

THEOREM 1. *If of all bang-bang steering functions there is an optimal one (relative to Ω^0), then it is optimal (relative to Ω).*

THEOREM 2. *If there is an optimal steering function (in Ω) then there is always a bang-bang steering function (in Ω^0) that is optimal.*

From Lemma 1 it is also not difficult to show that

THEOREM 3. *If for the general problem there is a steering function u in Ω such that $x(t, u) = z(t)$ for some $t > 0$, then there is an optimal steering function in Ω. Moreover, all optimal steering functions u* are of the form*

$$u^*(t) = \text{sgn}[\eta Y(t)] \qquad (5)$$

where η is some n-dimensional vector. (For r-dimensional vectors a and b, a = sgn b means that $a_i = \text{sgn } b_i, i = 1, \ldots, r$.)

Let $y^j(t)$ be the jth column vector of $Y(t)$. The control system (1) is said to be *normal* if on each interval of positive length and for each $j = 1, \ldots, r$ the functions $y_1^j(t), \ldots, y_n^j(t)$ are linearly independent. This is equivalent to saying that no component of $\eta Y(t), \eta \neq 0$, is identically zero on an interval of positive length, and therefore $u^*(t)$ is uniquely determined by (5). Hence

THEOREM 4. *For normal control systems the general problem has at most one optimal steering function.*

Thus the only way of reaching the objective in minimum time using a normal system is by at all times utilizing properly all of the power available.

3.   The Special Problem.—The control problem for the system

$$\dot{x}(t) = A(t)x(t) + B(t)u(t), \tag{6}$$

where the objective is to start at the initial state x and to reach the origin (the equilibrium state) in minimum time will be called *the special problem*. Hence for the special problem we want (see equation (4))

$$-x_0 = \int_0^t Y(\tau)u(\tau)d\tau. \tag{7}$$

It is then not difficult to show that

Theorem 5. *If for some $t > 0$ and some n-vector η there is a solution $u = u^*$ of (7) of the form*

$$u^*(\tau) = \operatorname{sgn}[\eta Y(\tau)], \tag{5}$$

then it is an optimal steering function for the special problem.

It is this result that is of interest in solving the synthesis problem. If the control system is autonomous (equation (2)), then we can start the control system at the origin, use a steering function of the form (4) and look at the solution as t decreases (replace t by $-t$). This steering function is then optimal for all the states that can be reached in this manner. Reversing the system in this way gives the set of all initial states in the special problem for which this steering function is optimal. For normal systems the optimal steering is unique, and this procedure always determines the optimal steering as a function of the state of the system. We say "always" in the above sentence because we know that the synthesis problem can be solved in this way for some systems that are not normal. This procedure leads to the determination of switching-surfaces, which are surfaces where certain of the components of the steering change sign.

It is now that we can see the usefulness of introducing another classification of control systems. If $\eta Y(t) \equiv 0$ on any interval of positive length implies $\eta = 0$, then the control system (1) is said to be *proper*. This is equivalent to saying that the row vectors $y_1(t), \ldots, y_n(t)$ of $Y(t)$ are linearly independent vector functions on each interval of positive length. It is clear that *every normal control system is proper* but the converse is not true. It is also not difficult to see, when we remove all constraints on the admissible control functions, that *proper control systems are completely controllable,*[7] i.e., given any two states x_1 and x_2 and any two times t_1 and $t_2, t_1 \neq t_2$, there is a steering function such that starting at x_1 at time t_1 the system is brought to the state x_2 at time t_2.

Proper systems also have the additional controllability property (now we return to the constraint $|u_i(t)| \leq 1$):

THEOREM 6. *If the system (2) is proper and asymptotically stable* $(X(t) \to 0$ as $t \to \infty)$, *then for each initial state* x_0 *there is a steering function in* Ω *that brings the system to the origin in minimum time.*

It is easy to see for proper systems that optimal steering functions lie on the boundary of Ω. Expressed as a bang-bang principle this states that: *In proper control systems optimal steering* u^* *has the property that at any given time some component of* u^* *is utilizing the maximum power available to it.*

It is of considerable importance to observe that for proper control systems there is a way (if the optimal system for the special problem can be synthesized) of deciding whether or not it is possible to start at a point x_0 and then hit the moving particle $z(t)$ and also possible to determine optimal steering. We can state this result as follows:

For proper control systems the problem $x(0, u) = x_0$, $x(t, u) = z(t)$ *for some* $t > 0$ *and some* u *in* Ω, *has a solution if and only if it is possible to start at some point* $-w(t_1)$ *and then with steering in* Ω *to reach the origin in time* $t_2 \leq t_1$. *If* $-w(t_1)$ *is the first point on the curve* $-w(t)$, $t > 0$, *from which it is possible to reach in this manner the origin in time* t_1, *then any steering that does this is optimal for this special problem and is also optimal for the general problem of hitting* $z(t)$.

* This research was supported in part by the Air Force Office of Scientific Research under Air Force Contract AF 49 (638)-382.

[1] Bushaw, D. W., Ph.D. Thesis, Department of Mathematics, Princeton University, 1952: "Differential Equations with a Discontinuous Forcing Term, Experimental Towing Tank," Stevens Institute of Technology Report No. 469 (January, 1953); "Optimal Discontinuous Forcing Terms," Contributions to the Theory of Nonlinear Oscillations, IV, Princeton, 1958.

[2] LaSalle, J. P., Abstract 247t, *Bull. Am. Math. Soc.*, **60**, 154 (1954); Study of the Basic Principle Underlying the "Bang-Bang" Servo, Goodyear Aircraft Corp. Report GER-5518 (July, 1953).

[3] Bellman, R., I. Glicksberg, and O. Gross, "On the 'Bang-Bang' Control Problem," *Q. Appl. Math.*, **14**, 11–18 (1956).

[4] Krasovskii, N. N., "Concerning the Theory of Optimal Control," *Avtomat. i Telemeh.*, 960–970 **18**, (1957). (Russian.)

[5] Gamkrelidze, R. V., "Theory of Time-Optimal Processes for Linear Systems," *Izvestia Akad. Nauk, SSSR*, **22**, 449–474 (1958). (Russian.)

[6] A simple and elegant proof of this Lemma was pointed out to me by L. Pukanszki using a theorem of Liapunov. See Liapunov, A., "Sur les fonctions-vecteurs complètement additives," *Bull. Acad. Sci. URSS Ser. Math. Izvestia Akad. Nauk, SSSR*, **4**, 465–478 (1940); Halmos, P. R., "The Range of a Vector Measure," *Bull. Am. Math. Soc.*, **54**, 416–421 (1948).

[7] The concept of controllability has been introduced by R. E. Kalman and will be discussed by him in a paper to appear.

ON THE THEORY OF OPTIMAL PROCESSES*

by V. Boltyanskii, R. Gamkrelidze and L. Pontryagin

CONSIDER the problem of determining the control function y in such a way that the system whose motion is described by the equations

$$\frac{dx}{dt} = g(x, y), \qquad x(0) = c \tag{1}$$

is transformed into the null state as rapidly as possible, where the function y is constrained to be a member of a set U. If we let $f(c)$ represent the minimum time, then, using the principle of optimality, we find that $f(c)$ satisfies the equation

$$-1 = \min_{y \in U} \{g \cdot \operatorname{grad} f\}. \tag{2}$$

At each point in the phase space of state vectors c the control vector y is to be chosen so as to maximize the scalar product of the given vector g and the unknown vector $\operatorname{grad} f = (\partial f/\partial c_1, \ldots, \partial f/\partial c_n)$. The functional equation approach of dynamic programming leads to the equation (2) from which we can determine the function $f(c)$ for all state vectors c of interest and, of course, determine optimal choices of the control y for each state vector c. Since numerical integration must ordinarily be considered we face all the difficulties of dealing numerically with functions of n variables. Although reduction of dimensionality using linearity, Lagrange multipliers, and polynomial approximation can often be effected, this straightforward approach possesses difficulties; see:

R. Bellman and S. Dreyfus, *Applied Dynamic Programming*. Princeton: Princeton University Press, 1962.

R. Bellman, R. Kalaba, and B. Kotkin, "Polynomial Approximation—A New Computational Technique in Dynamic Programming: Allocation Processes," *Mathematics of Computation*, Vol. 17, 1963, pp. 155–161.

* Translated by J. H. Jones for the Technical Library, Space Technology Laboratories, Inc., from *Reports of the Academy of Sciences of the USSR*, Vol. 110, No. 1, 1956, pp. 7–10.

Another view consists in attempting to find the function grad f not at all points, but only at points along an optimal trajectory from an initial point c to the given terminal point. When this is known, by optimizing $g \cdot \mathrm{grad}\, f$ with respect to y at each point, the optimal control is determined.

In the paper that follows, the extremely elegant maximum principle of Pontryagin is introduced to accomplish this objective. In many cases, such as the "bang-bang" control process, it provides a very simple and direct means of determining the nature of the optimal control policy. The difficulty here, from the numerical view, is that a two-point boundary value problem for a system of nonlinear differential equations results. The method of quasilinearization is useful in this conjunction; see:

R. Kalaba, "On Nonlinear Differential Equations, the Maximum Operation, and Monotone Convergence," *J. Math. and Mech.*, Vol. 8, 1959, pp. 519–574.

Various connections between the work of Pontryagin and his school to the classical work of Caratheodory and others has been pointed out by L. Berkovitz; see:

L. Berkovitz, "Variational Methods in Problems of Control and Programming," *J. Math. Anal. and Appl.*, Vol. 3, 1961, pp. 145–169.

See also:

C. Desoer, "Pontriagin's Maximum Principle and the Principle of Optimality," *J. Franklin Institute*, May 1961, pp. 361–367.

R. Rozonoer, "L. S. Pontryagin Maximum Principle in the Theory of Optimal Systems I, II, III," *Automation and Remote Control*, Vol. 20, 1959, pp. 1288–1302, etc.

E. O. Roxin, "Geometric Interpretation of Pontryagin's Maximum Principle," *Nonlinear Differential Equations and Nonlinear Mechanics*. New York: Academic Press, Inc., 1962.

Finally, for a detailed account, see:

L. S. Pontryagin, V. G. Boltyanskii, R. V. Gamkrelidze, and E. F. Mishchenko, *Mathematical Theory of Optimal Processes*. Moscow: 1961. (English translation, New York: (Interscience) John Wiley and Sons, 1962.)

On the Theory of Optimal Processes

V. G. BOLTYANSKII, R. V. GAMKRELIDZE,

AND L. S. PONTRYAGIN

Translated by J. H. Jones
for the Technical Library, Space Technology Laboratories, Inc.

IN recent years, in the theory of automatic control, great importance has been attached to the most rapid realization of control processes, which has led to the appearance of a series of works devoted to the study of so-called *optimal processes* (see (^{1}) for a bibliography list). In this paper we consider the general approach to the study of optimal processes.

1. *Statement of the Problem.*—Let us consider a representative point $(x^1, \ldots, x^n) = x$ in n-dimensional phase space, for which the equations of motion are given in the normal form

$$\dot{x}^i = f^i(x^1, \ldots, x^n; u^1, \ldots, u^r) = f^i(x, u), \qquad i = 1, \ldots, n. \qquad (1)$$

Here $u^1, \ldots, u^r$ are the *control parameters*. If the control rule is given, i.e. the variable vector $u(t) = (u^1(t), \ldots, u^r(t))$ is given in r-dimensional space, then the system (1) uniquely determines the motion of the point.

We impose the natural conditions of piece-wise smoothness and piece-wise continuity on "the steering vector" $u(t)$. In addition, we will assume that the variable vector $u(t)$ belongs to a fixed closed region $\overline{\Omega}$ in the space of the variables $u^1, \ldots, u^r$. This is the closure of the open region Ω with piece-wise smooth, $(r - 1)$-dimensional, boundary. For example, the region $\overline{\Omega}$ may be an r-dimensional cube: $|u^i| \leq 1$, $i = 1, \ldots, r$; the half-space $u^i \geq 0$, etc. We shall call the control vector $u(t)$, which satisfies the enumerated conditions, *admissible*.

Formulation of the General Problem.—Given two points ξ_0, ξ_1 in the phase space $x^1, \ldots, x^n$; it is required to choose the admissible control vector $u(t)$ such that the point passes from the position ξ_0 to the position ξ_1 in minimal time.

The steering vector $u(t)$ which we are trying to find will be called the *optimal control*. The corresponding trajectory $x(t) = (x^1(t), \ldots, x^r(t))$ of system (1) will be called the optimal trajectory.

2. *Necessary Conditions for Optimalization.*—Let us assume that the optimal $u(t)$ and the optimal trajectory corresponding to it exist. The

trajectory $x(t)$ satisfies the boundary conditions $x(t_0) = \xi_0$, $x(t_1) = \xi_1$. Let us first assume that the directing vector $u(t)$ for $t_0 = t = t_1$ lies inside the open region Ω. Consequently, for any sufficiently small perturbation (with respect to the modulus of the perturbations $\delta u(t) = (\delta u^1(t), \ldots, \delta u^r(t))$ of the vector $u(t)$), the direction $u(t) + \delta u(t)$ will remain in region Ω. By $x + \delta x$, we mean the "perturbation" (corresponding to the control $u(t) + \delta u(t)$) of the trajectory, with the same initial condition $x(t_0) + \delta x(t_0) = \xi_0$, i.e. $\delta x(t_0) = 0$. The equations of the linear approximation

$$\delta_1 x = (\delta_1 x^1, \ldots, \delta_1 x^n)$$

for the perturbation $\delta x = (\delta x^1, \ldots, \delta x^n)$ has the form

$$\delta_1 \dot{x}^1 = \frac{\partial f^i}{\partial x^\alpha} \delta_1 x^\alpha + \frac{\partial f^i}{\partial u^\beta} \delta u^\beta, \quad \delta_1 x(t_0) = 0; \quad i = 1, \ldots, n. \tag{2}$$

Because of the linearity of system (2), the points $x(t_1) + \delta_1 x(t_1)$ (which correspond to all possible perturbations $\delta_1 u(t)$, which are sufficiently small with respect to the modulus) fill up a region of some linear manifold P', passing through the point $x(t_1)$. From the optimal property of the trajectory $x(t)$ it is easily seen that the dimension of P' does not exceed $n - 1$, and P', generally speaking, is not tangent to the trajectory $x(t)$. Let $P(t_1)$ be an $(n - 1)$-dimensional plane containing P', and not tangent to the trajectory $x(t)$. We denote the covariant coordinates of the plane $P(t_1)$ by $a_1, \ldots, a_n$; then $a_\alpha \delta_1 x^\alpha(t_1) = 0$.

Let $\phi_j(t) = (\phi_j^1(t), \ldots, \phi_j^n(t))$, $(j = 1, \ldots, n)$, be the fundamental system of solutions of the homogeneous system corresponding to system (2), and let $\|\psi_j^i(t)\|$ be the inverse matrix of the matrix $\|\phi_j^i(t)\|$. The solution of the system (2) may be written in the form

$$\delta_1 x^i(t) = \phi_\alpha^i(t) \int_{t_0}^t \psi_\beta^\alpha \frac{\partial f^\beta}{\partial u^\gamma} \delta u^\gamma \, d\tau, \quad i = 1, \ldots, n. \tag{3}$$

Using the relation $a_\alpha \delta_1 x^\alpha(t_1) = 0$, we obtain

$$a_\alpha \delta_1 x^\alpha(t_1) = a_\alpha \phi_\beta^\alpha(t_1) \int_{t_0}^{t_1} \psi_\gamma^\beta \frac{\partial f^\gamma}{\partial u^\nu} \delta u^\nu \, d\tau = 0.$$

Let us introduce the transformations:

$$a_\alpha \phi_\beta^\alpha(t_1) = b_\beta, \quad b_\beta \psi_\gamma^\beta(t) = \psi_\gamma(t).$$

Then,

$$a_\alpha \delta_1 x^\alpha(t_1) = \int_{t_0}^{t_1} \psi_\alpha \frac{\partial f^\alpha}{\partial u^\beta} \delta u^\beta \, d\tau = 0.$$

Since $\delta u(t) = (\delta u^1(t), \ldots, \delta u^r(t))$ is an arbitrary perturbation (sufficiently small with respect to the modulus), the last equation implies that the system of equations

$$\psi_\alpha(t) \frac{\partial f^\alpha}{\partial u^i} = 0, \quad t_0 \leq t \leq t_1, \quad i = 1, \ldots, r. \tag{4}$$

The vector $\psi(t) = \psi_1(t), \ldots, \psi_n(t))$ has a simple geometric interpretation: the point $x(t) + \delta_1 x(t)$ lies in the $(n-1)$-dimensional plane $P(t)$, which passes through the point $x(t)$, and which has covariant coordinates $\psi_1(t), \ldots, \psi_n(t)$. In particular, $(\psi_1(t_1), \ldots, \psi_n(t_1)) = (a_1, \ldots, a_n)$. Using the relation $\psi_i(t) = b_\alpha \psi_i^\alpha(t)$, $i = 1, \ldots, n$, we can deduce the following system of differential equations for $\psi_i(t)$:

$$\psi_i(t) = -\frac{\partial f^\alpha}{\partial x^i}\psi_\alpha, \qquad i = 1, \ldots, n. \tag{5}$$

Combining systems (1), (4), (5):

$$\left.\begin{aligned}
\dot{x}^i &= f^i(x, u), \quad i = 1, \ldots, n; \\
\dot{\psi}_i &= -\frac{\partial f^\alpha}{\partial x^i}\psi_\alpha, \quad i = 1, \ldots, n; \\
\psi_\alpha \frac{\partial f^\alpha}{\partial u^j} &= 0, \quad t_0 \le t \le t_1, \quad j = 1, \ldots, r.
\end{aligned}\right\} \tag{6}$$

System (6) represents a set of necessary conditions which the optimal control $u(t)$ (if it is to lie inside the open region Ω), its corresponding optimal trajectory $x(t)$, and the vector $\psi(t)$ must satisfy.

Multiplying the vector $\psi(t)$ by a suitable constant (which does not change the trajectory $x(t)$ by the control $u(t)$), we obtain $\psi_\alpha(t_0)f^\alpha(x(t_0), u(t_0)) > 0$. Since the plane $P(t)$ is not tangent to the trajectory $x(t)$, i.e. $\psi_\alpha f^\alpha \ne 0$ for any t, the inequality $\psi_\alpha f^\alpha > 0$ will be satisfied for every t.

If we now assume that the optimal control may belong to the *closed* region $\bar\Omega$ and take the inequality $\psi_\alpha f^\alpha|_{t=t_0} > 0$ into account, then system (4) is replaced by the more general condition

$$\psi_\alpha \frac{\partial f^\alpha}{\partial u^\beta}\delta u^\beta \le 0, \qquad t_0 \le t \le t_1, \tag{7}$$

for arbitrary perturbations $\delta u^\beta(t)$, constrained by the condition

$$u(t) + \delta u(t) \in \bar\Omega.$$

3. *Sufficient Conditions for Optimalization in the Small.*—In this section we again assume that the steering vector $u(t)$ lies inside the open region Ω and consequently that the necessary conditions (6) are fulfilled. The equations for the second order approximation $\delta_{11}x$ of the perturbation δx have the form:

$$\delta_{11}\dot{x}^1 = \frac{\partial f^i}{\partial x^\alpha}\delta_{11}x^\alpha + \frac{\partial f^i}{\partial u^\beta}\delta u^\beta + B^i(t),$$

where

$$B^i(t) = \frac{1}{2}\left[\frac{\partial^2 f^i}{\partial x^\alpha\,\partial x^\beta}\delta_1 x^\alpha\,\delta_1 x^\beta + 2\frac{\partial^2 f^i}{\partial x^\alpha\,\partial u^\beta}\delta_1 x^\alpha\,\delta u^\beta + \frac{\partial^2 f^i}{\partial u^\alpha\,\partial u^\beta}\delta u^\alpha\,\delta u^\beta\right].$$

The point with coordinates

$$x^i(t) + \delta_{11}x^i(t) = x^i(t) + \delta_1 x^i(t) + \phi^i_\alpha(t) \int_{t_0}^t \psi^\alpha_\beta B^\beta \, d\tau$$

does not lie in the plane $P(t)$. If the point moving with the perturbed motion, at some time t, passed beyond the plane $P(t)$, then the scalar product

$$\psi_\alpha(t) \, \delta_{11}x^\alpha(t) = \psi_\alpha(t) \, \delta_1 x^\alpha(t) + \psi_\alpha(t)\phi^\alpha_\beta(t) \int_{t_0}^t \psi^\beta_\gamma B^\gamma \, d\tau$$

$$= \psi_\alpha(t)\phi^\alpha_\beta(t) \int_{t_0}^t \psi^\beta_\gamma B^\gamma \, d\tau = \int_{t_0}^t \psi_\alpha B^\alpha \, d\tau$$

would be positive. If the point did not go as far as the plane $P(t)$, then,

$$\psi_\alpha(t) \, \delta_{11}x^\alpha(t) = \int_{t_0}^t \psi_\alpha B^\alpha \, d\tau < 0.$$

Let the quadratic form

$$\psi_\alpha \frac{\partial^2 f^\alpha}{\partial u^i \, \partial u^k} \, \delta u^i \, \delta u^k$$

(of the variables $\delta u^1, \ldots, \delta u^r$) be negative-definite at the point $(x(t_0), u(t_0), t_0)$. Then the scalar product

$$\psi_\alpha(t) \, \delta_{11}x^\alpha(t) = \int_{t_0}^t \psi_\alpha B^\alpha \, d\tau \leq 0$$

for any sufficiently small (with respect to the modulus) perturbations $\delta u(t)$, and for $t - t_0$ sufficiently small. In this case the control $u(t)$ and the trajectory $x(t)$ are *optimal in the small*, i.e. there is a neighborhood V about the point $x(t_0)$ such that if $x(t')$, $x(t'')$ ($t' < t''$) are arbitrary points of the trajectory, belonging to V, then there is no control, sufficiently close to $u(t)$, which makes it possible to get from the point $x(t')$ to $x(t'')$ in a shorter time than $t'' - t'$.

If the form

$$\psi_\alpha \frac{\partial^2 f^\alpha}{\partial u^i \, \partial u^k} \, \delta u^i \, \delta u^k,$$

at the point $(x(t_0), u(t_0), t_0)$, is not definite, then (with some sufficiently general additional condition) no control vector $u(t)$, which belongs to the open region Ω near $t = t_0$, can be optimal, even in the small. If optimal trajectories passing through the point $x(t_0)$ nevertheless exist, then the corresponding control vector $u(t)$, near $t = t_0$, must lie on the boundary of the closed region $\overline{\Omega}$.

4. *The Principle of the Maximum.*—From system (6) and the negative definiteness of the quadratic form

$$\psi_\alpha \frac{\partial^2 f^\alpha}{\partial u^i \, \partial u^k} \, \delta u^i \, \delta u^k,$$

it follows that the expression $\psi_\alpha(t)f^\alpha(x(t),\,u(t))$ achieves, for fixed vectors $x(t)$, $\psi(t)$, and variable vector $u(t)$, a relative maximum: for sufficiently small (with respect to the modulus) perturbations $\delta u(t)$ the following inequality holds:

$$\psi_\alpha(t)f^\alpha(x(t),\,u(t)) \geq \psi_\alpha(t)f^\alpha(x(t),\,u(t) + \delta u(t))$$

for all time, as long as equation (6) is satisfied and the quadratic form remains negative definite.

This fact is a special case of the following general principle, which we call the principle of the maximum (this principle has been shown for some special cases only):

Let the function $H(x,\,\psi,\,u) = \psi_\alpha f^\alpha(x,\,u)$, *for any fixed* x, ψ *have a maximum with respect to* u, *when the vector* u *is varied in the closed region* $\overline{\Omega}$. *We denote this maximum by* $M(x,\,\psi)$. *If the $2n$-dimensional vector* $(x,\,\psi)$ *is the solution to the Hamiltonian system*

$$\left.\begin{aligned}
\dot{x}^i &= f^i(x,\,u) = \frac{\partial H}{\partial \psi_i}, \\
\dot{\psi}_i &= -\frac{\partial f^\alpha}{\partial x^i}\psi_\alpha = -\frac{\partial H}{\partial x^i},
\end{aligned}\right\} \quad i = 1,\ldots,n, \tag{8}$$

where the piecewise continuous vector $u(t)$ *satisfies the condition*

$$H(x(t),\,\psi(t),\,u(t)) = M(x(t),\,\psi(t)) > 0$$

for all t, *then* $u(t)$ *is an optimal control, and* $x(t)$ *corresponds to the optimal trajectory (in the small) of the system (1).*

Let the fixed initial condition $x(t_0) = \xi_0$ *be given, and for all possible forms let us vary the initial condition* $\psi(t_0) = \eta_0$. *Then system (8), together with these initial conditions, and the condition*

$$H(x(t),\,\psi(t),\,u(t)) = M(x(t),\,\psi(t)) > 0$$

determines the set of all optimal (in the small) trajectories which pass through the point $x(t_0) = \xi_0$, *with corresponding optimal control* $u(t)$.

Quoted Literature

1. A. A. Fel'dbaum, "The Second All-Union Conference on Automatic Control," Vol. 2, 1955, p. 325.

ON THE APPLICATION OF THE THEORY OF DYNAMIC PROGRAMMING TO THE STUDY OF CONTROL PROCESSES*

by Richard Bellman

DYNAMIC programming is a mathematical theory of multistage decision processes. It was developed to overcome or circumvent a number of formidable conceptual, analytic, and computational difficulties arising in contemporary optimization and control processes. More generally, it provides a means of treating problems which can be cast in the mold of sequential processes. Since the calculus of variations can be considered to be a multistage decision process of continuous type, dynamic programming can be used to derive a number of the basic analytic results; see, for example:

S. Dreyfus, "Dynamic Programming and the Calculus of Variations," *J. Math. Anal. and Appl.*, Vol. 1, 1960, pp. 228–239,

and to provide numerical algorithms, see:

R. Bellman and S. Dreyfus, *Applied Dynamic Programming*. Princeton: Princeton University Press, 1962.

A general problem which may be approached by these means is that of minimizing the functional

$$J(y) = \int_0^T g(x, y) \, ds, \tag{1}$$

where the vectors x and y are related by means of the differential equation

$$\frac{dx}{dt} = h(x, y), \qquad x(0) = c, \tag{2}$$

and subject to some further inequalities which may be algebraic or differential in nature. Here y is the control vector and x is the state vector.

A problem of greater difficulty in many ways is that where stochastic elements are present. There are now several distinct ways of defining optimal policies, dependent upon the type of information that is available at

---

* From *Proceedings of the Symposium on Nonlinear Circuit Analysis*, Polytechnic Institute of Brooklyn, Polytechnic Press, 1956, pp. 199–213.

each stage of the process, and the variety of control allowable. In the paper that follows, it is shown that dynamic programming can be used to provide straightforward formulation of stochastic feedback control processes, completely analogous to that given for deterministic processes, and just as readily used for computational and analytic solutions. Further results and references will be found in:

R. Bellman, *Adaptive Control Processes: A Guided Tour*. Princeton: Princeton University Press, 1961.

It is interesting to point out that the format of stochastic decision processes may be used to provide a more general and realistic formulation of the capacity of communication channels than that provided by "information theory." This approach was started by J. L. Kelly and extended in:

R. Bellman and R. Kalaba, "On Communication Processes Involving Learning and Random Duration," *1958 IRE National Convention Record*, Part 4, pp. 16–20.

Independently, these ideas were put forth by J. Marschak.
For applications to sequential analysis and detection theory, see:

R. Bellman, R. Kalaba, and D. Middleton, "Dynamic Programming, Sequential Estimation, and Sequential Detection Processes," *Proc. Nat. Acad. Sci. USA*, Vol. 47, 1961, pp. 338–341.

ON THE APPLICATION OF THE THEORY OF DYNAMIC
PROGRAMMING TO THE STUDY OF CONTROL PROCESSES[*]

Richard Bellman
The Rand Corporation

1. Introduction

In this paper we wish to discuss the application of the theory of
dynamic programming to the study of some representative control
processes of the type that arise in servomechanism theory and other
parts of engineering analysis, electrical and mechanical, as well.

We shall first formulate a general class of control problems,
in abstract fashion, and then present a number of illustrative examples, of both deterministic and stochastic type. Among these will be
the "bang-bang" control process, which has been extensively studied
in recent years.[14]

In order to make the paper reasonably self-contained, we shall
discuss the fundamentals of the theory of dynamic programming before presenting the applications to the calculus of variations in connection with specific processes. We shall bypass any analytical study
of the processes, and turn directly to the computational solution of
some typical control problems.

The purpose of this paper is to present a simple method, requiring a minimal mathematical background, which can be used to
treat a large class of variational problems, without regard to linear
or nonlinear, deterministic or stochastic features of the underlying
process.

The processes discussed here are analytically equivalent to a
number of allocation processes arising in mathematical economics,
and actually equivalent to "smoothing processes" arising in industrial
production.[5,9]

2. An Abstract Formulation of Control Processes

Let S denote a physical system whose state at any time t we
assume to be completely specified by an n-dimensional vector $x(t)$.
Assume further that $x(t)$ is determined for all $t \geq 0$ as a solution
of a functional equation of the form

$$L(x) = 0 , \qquad (2.1)$$

[*]Presented at the Symposium on Nonlinear Circuit Analysis,
Polytechnic Institute of Brooklyn, April 25 - 27, 1956.

together with appropriate initial conditions.

The two cases of greatest importance are those where $L(x)$ represents a differential operator, so that (2.1) assumes the form

$$\frac{dx}{dt} = g(x), \qquad x(0) = c, \tag{2.2}$$

where $g(x)$ is a vector function of the vector x, and where

$$x(t + 1) = g\left[x(t)\right], \quad x(0) = c, \tag{2.3}$$

with t assuming only a discrete set of values, $t = 0, 1, 2, \ldots$.

A more general equation, including both of the above as special cases, is the differential-difference equation,

$$\frac{dx}{dt} = g\left[x(t), x(t-1)\right], \tag{2.4}$$

with an initial condition

$$x(t) = a(t), \qquad -1 \leqslant t < 0. \tag{2.5}$$

Equations of this last type, together with a number of more complicated forms, arise upon taking account of time-lags and retarded control. A discussion of the theory and origin of these equations may be found in Reference 3.

Since the study of control processes in this field has been only briefly begun, we shall restrain our attention in this paper to processes described by either Eq. (2.2) or Eq. (2.3).

Consider then a system whose intrinsic equation is that of (2.2), and suppose that $g(x)$ satisfies appropriate conditions ensuring existence and uniqueness of the solution for $t \geqslant 0;$[2] It may happen that $x(t)$ is not a desirable state for the system, in the sense that we prefer it to be in the state represented by $y(t)$. Furthermore, we have a measure of our discomfort, a norm $\|x - y\|$, measuring the deviation of the actual state $x(t)$ from the desired state.

Of the many courses of action at our disposal for reducing the measure of displeasure, $\|x - y\|$, let us choose only one. We shall suppose that we can introduce an external control which manifests itself mathematically by way of a forcing term, with the result that (2.2) takes the form

$$\frac{dx}{dt} = g(x) + f(t), \qquad x(0) = c. \tag{2.6}$$

The norm, $\|x - y\|$, will now depend upon $f(t)$, which is to say that it will be a function of $f(t)$. The general control problem is then to choose $f(t)$, subject to certain feasibility constraints, so as to minimize

$$J(f) = ||x - y|| . \qquad (2.7)$$

3. Some Typical Problems

Retreating from this lofty plane of abstraction, let us consider some specific control processes.

The equation,

$$x'' + \lambda(x^2 - 1) x' + x = 0, \qquad x(0) = c_1, \qquad x'(0) = c_2, \qquad \lambda > 0, \qquad (3.1)$$

(the famed equation of Van der Pol) arising in the study of a multivibrator, possesses, as we know, a single periodic solution to which all solutions tend, regardless of their initial values.[*]

Suppose that in some particular process this oscillation is actually parasitic, and that we would like the system described by (3.1) to remain in the equilibrium position,

$$x = 0, \qquad x' = 0 . \qquad (3.2)$$

To rid ourselves of this unwanted oscillation, we introduce a control term in the form of a forcing term $f(t)$. The equation describing the system now has the form

$$x'' + \lambda(x^2 - 1) x' + x = f(t), \qquad x(0) = c_1, \qquad x'(0) = c_2. \qquad (3.3)$$

We wish to choose $f(t)$, subject to constraints imposed by the physical nature of the system, so as to reduce x to its equilibrium state and maintain it there, or to minimize the deviation of x from its desired state.

As a first approach to this problem, we may consider the problem of minimizing

$$J(x) = \int_0^T (x^2 + x'^2) \, dt, \qquad (3.4)$$

a measure of the deviation of the system from equilibrium, subject to a constraint of the form

$$\int_0^T f^2 \, dt \le k_1 \qquad (3.5)$$

a measure of the cost of control.

[*]Apart from $c_1 = c_2 = 0$, of course.

202   NONLINEAR CIRCUIT ANALYSIS

Using a Lagrange multiplier a_1, this is equivalent to the problem of minimizing

$$J(f) = \int_0^T (x^2 + x'^2 + a_1 f^2)\, dt, \qquad (3.6)$$

or

$$J(x) = \int_0^T \left[x^2 + x'^2 + a_1 (x'' + \lambda(x^2 - 1)x' + x)^2 \right] dt. \qquad (3.7)$$

At this point, we may brighten up, feeling that the problem is safely within the orbit of the calculus of variations. Let us, however, pursue the solution a few steps further before attempting to dismiss it as routine. Writing

$$J(x) = \int_0^T F(x, x', x'')\, dt, \qquad (3.8)$$

let x denote a desired minimizing function, assumed to exist,[*] and consider

$$g(\epsilon) = J(x + \epsilon y) = \int_0^T F(x + \epsilon y, x' + \epsilon y', x'' + \epsilon y'')\, dt, \qquad (3.9)$$

where ϵ is a real parameter and y an arbitrary function.

Since $g(\epsilon)$ possesses a minimum at $\epsilon = 0$, we must have

$$g'(0) = \frac{\partial}{\partial \epsilon} J(x + \epsilon y) \Big|_{\epsilon = 0}$$

$$= \int_0^T \left(y \frac{\partial F}{\partial x} + y' \frac{\partial F}{\partial x'} + y'' \frac{\partial F}{\partial x''} \right) dt \qquad (3.10)$$

$$= 0,$$

for all functions $y(t)$. Integrating by parts, we have

$$\int_0^T y \left[\frac{\partial F}{\partial x} - \frac{d}{dt} \frac{\partial F}{\partial x'} + \frac{d^2}{dt^2} \frac{\partial F}{\partial x''} \right] dt$$

$$(3.11)$$

$$+ \left[y' \frac{\partial F}{\partial x''} \right]_0^T + \left[y \left(\frac{\partial F}{\partial x'} - \frac{d}{dt} \frac{\partial F}{\partial x''} \right) \right]_0^T = 0$$

for all y satisfying the conditions $y(0) = 0$, $y'(0) = 0$.

---

[*]Let us agree to follow a purely formal approach for the moment. As we shall see, we can bypass a number of the thorny, rigorous details.

DYNAMIC PROGRAMMING   203

Hence we obtain the equation

$$\frac{\partial F}{\partial x} - \frac{d}{dt}\frac{\partial F}{\partial x'} + \frac{d^2}{dt^2}\frac{\partial F}{\partial x''} = 0, \qquad (3.12)$$

the Euler equation, together with the boundary conditions

$$\frac{\partial F}{\partial x''}\bigg|_{t=T} = 0,$$

$$\left[\frac{\partial F}{\partial x'} - \frac{d}{dt}\frac{\partial F}{\partial x''}\right]_{t=T} = 0. \qquad (3.13)$$

Equation (3.12) is a fourth order differential equation, requiring, in consequence, four boundary conditions to determine the solution. Since two conditions are furnished by the initial conditions,

$$x(0) = c_1, \qquad\qquad x'(0) = c_2, \qquad (3.14)$$

and two additional conditions furnished by the Eqs. (3.13), it appears as if the numerical solution is straightforward. Upon examining these four conditions more closely we observe that two conditions hold at $t = 0$, and the other two conditions at $t = T$. At neither point do we have a sufficient number of conditions to permit the solution to be obtained by a systematic point-by-point integration.

If Eq. (3.12) is nonlinear, as it is in the example above, we are confronted by a formidable two-point boundary-value problem. What is usually done is to take $x''(0)$ and $x'''(0)$ as unknowns and attempt to determine them as solutions of the equations in (3.13). This is a complicated set of simultaneous equations requiring a considerable amount of computation.

Assume, however, that we have resolved this problem. Nevertheless, we have no cause for smugness. Neither the measure of deviation, nor the measure of the cost of control, are satisfying in general. Suppose that we replace the constraint of (3.5) by

$$|f(t)| \leqslant k, \qquad (3.15)$$

in many cases a more realistic constraint.

In this case, graver difficulties arise. As long as the constraint in (3.15) is proper, i.e.,

$$-k < f(t) < k, \qquad (3.16)$$

we can employ a variational technique and obtain an Euler equation. However, there may also be, and generally will be, intervals where

$f(t) = k$, and $f(t) = -k$. The forcing function $f(t)$ may have the form shown in Fig. 1.

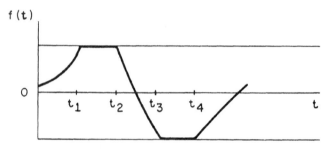

Figure 1

The problem of determining the transition points, t_i, at which one regime goes over into another, is one which at the present time can only be systematically resolved in certain simple cases.[10]

Assume, however, that we have resolved this problem. We are still not satisfied with the measure of deviation. The problem we would like to resolve is that of determining $f(t)$, subject to (3.15) so as to minimize

$$J(f) = \underset{0 \leqslant t \leqslant T}{\text{Max}} \left[x(t) \right]. \qquad (3.17)$$

At this point, classical techniques are stymied. We shall show below how this problem may be resolved numerically by means of the techniques of dynamic programming.

In place of the measure of deviation in (3.17), we may wish to determine $f(t)$, subject to the constraint of (3.15), so as to reduce x to the equilibrium state, $x = 0$, $x' = 0$, in a minimum time. This is a fundamental problem in servo-mechanism theory.[11,14] As we shall see, this problem may also be treated by means of functional equations. The approaches in the above references apply only to linear equations.

4. Dynamic Programming

The problems posed above demonstrate in a clear fashion that control processes give rise to a variety of variational questions, some of conventional type and some not. We wish to show that in a number of cases these variational problems can be resolved computationally in a simple, straightforward manner, with the aid of modern digital computers, when formulated as multi-stage decision processes.

Let us, then, briefly sketch the essentials of the theory of dynamic programming, leading up to the use of the functional equation technique. We shall consider a simple class of decision processes which include the types of control processes we wish to treat here.

DYNAMIC PROGRAMMING 205

More complete discussions of dynamic programming may be found in References 1 and 6.

Consider a physical system S whose state at any time can be characterized by an n-dimensional vector p, and let $\{T(p, q)\}$ denote a set of transformations which may be applied to S. Here q denotes a choice vector, specifying the transformation to be used. It may be an element of a finite set or an infinite set, and in many cases its range will depend upon p itself. If a particular member of the set of transformations, $T(p, q_1)$, is used, there are two effects:

(a) p is transformed into $p_1 = T(p, q_1)$,

(b) an "output", $r(p, q_1)$, a scalar function, is obtained.

Assume that q is to be chosen so as to maximize this output function. A process of this type we shall call a *single-stage decision process*.

Suppose now that the process consists of two stages, in the following manner. We first choose q_1, converting p into $p_1 = T(p, q_1)$, and then q_2, converting p_1 into $p_2 = T(p_1, q_2)$. From the first choice we obtain $r(p, q_1)$, and from the second choice $r(p_1, q_2)$. Let us assume that the purpose of this two-stage process is to maximize the total "output":

$$R_2 = r(p, q_1) + r(p_1, q_2). \qquad (4.1)$$

The vectors q_1 and q_2 will then be chosen so as to maximize this function of the variables q_1 and q_2.

Consider the general N-stage process. We choose $q_1, q_2, \ldots, q_N$, obtaining a sequence of state vectors,

$$p_1 = T(p, q_1)$$

$$p_2 = T(p_1, q_2)$$

$$\cdot$$
$$\cdot \qquad (4.2)$$
$$\cdot$$

$$p_N = T(p_{N-1}, q_N),$$

and a total "return"

$$R_N = r(p, q_1) + r(p_1, q_2) + \ldots + r(p_{N-1}, q_N). \qquad (4.3)$$

The set $(q_1, q_2, \ldots, q_N)$ is to be chosen so as to maximize R_N.

206   NONLINEAR CIRCUIT ANALYSIS

A process of this type is called a *multi-stage decision process*. [*]

As the problem stands, it seems like one which could profitably be treated by the methods of calculus. There are, however, two principal difficulties. In the first place, if the number of stages is large, the variational equations constitute a set of simultaneous equations in a large number of independent variables. The problem of solving this set of equations is, in the main, as formidable as the evaluation of the maximum directly by means of search techniques. Secondly, calculus may not be directly available due to the fact that the functions involved may not be sufficiently smooth, or, ironically, they may be too smooth, which is to say linear, forcing the maximum to boundary points of the region of variation, R.

We see then that there are more difficulties than might be apparent at first glance. We can, nonetheless, take advantage of the specific structure of this multi-stage process to lift a part of the curse of dimensionality.

The number of independent variables may be reduced in the following way. Observe that the maximum value of R_N depends only upon p, the initial vector, and N, the number of stages. Let us then define the auxiliary sequence of functions,

$$f_N(p) = \underset{\{q_i\}}{\text{Max}}\ R_N, \qquad (4.4)$$

for $N = 1, 2, \ldots$, and all p in R. We shall assume that $p \in R$ implies that $T(p, q) \in R$ for all admissible q, so that we may continually consider a fixed region.

Having thus defined the maximization process for all N and for all p, we observe that the process possesses a certain important *invariant* property. After any initial choice of q_1, the remaining $(N - 1)$ stages of the process constitute, in themselves, an $(N - 1)$ stage decision process, starting from the new state $T(p, q_1)$. [**]

Using this remark concerning the invariant nature of the process, we can obtain a recurrence relation connecting two consecutive members of the sequence, $f_N(p)$. Consider an N-stage process, with an initial choice of q_1. By virtue of the above, the remaining choices $(q_2, q_3, \ldots, q_N)$ must be made so as to yield a total return of $f_{N-1}\left[T(p, q_1)\right]$ from the final $(N - 1)$ stages of the process.

Thus, whatever the initial choice of q_1, we will have

$$R_N = r(p, q_1) + f_{N-1}\left[T(p, q_1)\right], \qquad (4.5)$$

---

*As we have mentioned above, this is actually a particular type of multi-stage decision process, covering only the class of problems we treat in Sections 1-5. For a more general definition see Reference 6.

**This is an application of the 'principle of optimality'. [1,6]

if we wish to maximize R_N. Since q_1 itself must be chosen to maximize, we obtain the recurrence relation

$$f_N(p) = \underset{q_1}{\text{Max}} \left[r(p, q_1) + f_{N-1} \left[T(p, q_1) \right] \right] , \qquad (4.6)$$

for $N = 2, 3, \ldots$, with

$$f_1(p) = \underset{q_1}{\text{Max}} \left[r(p, q_1) \right] . \qquad (4.7)$$

The importance of the above relations lies in the fact that we
have replaced the original problem, involving an N n-dimensional
problem, N stages × n dimensions, by a sequence of N n-dimensional problems. Since the only operation involved in computing the
various members of the sequence is that of determining a maximum,
there is no necessity for choosing simple analytic forms for $r(p, q)$
and $T(p, q)$. It is this property which makes this method particularly
applicable to problems which escape any classical approach because
of non-analytic structure of $r(p, q)$ and $T(p, q)$.

5.   Variational Processes as Multi-Stage Decision Processes
      Let us now indicate how an important class of variational proc-
esses may be regarded as multi-stage decision processes of the above
type. Consider, to begin with, the problem of determining the maximum of the integral

$$J(y) = \int_0^T F(x, y) \, dt, \qquad (5.1)$$

over all functions $y(t)$, where $x(t)$ and $y(t)$ are scalar variables
connected by the relation

$$\frac{dx}{dt} = G(x, y), \qquad x(0) = c. \qquad (5.2)$$

The special case, $G(x, y) \equiv y$, yields the familiar problem of maximizing

$$J(x) = \int_0^T F(x, \frac{dx}{dt}) \, dt. \qquad (5.3)$$

The first step in our conversion of variational problems to
multi-stage decision processes lies in the consideration of the discrete version of the above continuous problem. Although we can treat
the continuous process directly,[4, 6] the method used here seems
better adapted for computational purposes. It should be remembered
in any case that a digital computer converts all processes into discrete

208   NONLINEAR CIRCUIT ANALYSIS

processes, whatever their origin.

Let the time interval $[0, T]$ be divided into N subintervals of length Δ,

$$\vdash\!\!\!\!\!-\!\!\!\!\!-\!\!\!\!\!-\!\!\!\!\!-\!\!\!\!\!-\!\!\!\!\!-\!\!\!\!\!-\!\!\!\!\!-\!\!\!\!\!-\!\!\!\!\!-\!\!\!\!\!-\!\!\!\!\!-\!\!\!\!\!-\!\!\!\!\!-\!\!\!\!\!-\!\!\!\!\!-\dashv$$

0   Δ 2Δ $(N-1)\Delta$ $N\Delta = T$

In place of considering all functions $y(t)$, defined for $0 \leqslant t \leqslant T$, we consider all sequences, $\{y(k\Delta)\}$, $k = 0, 1, 2, \ldots, N$. The integral $J(y)$ is replaced by the sum

$$J(\{y_k\}) = \sum_{k=0}^{N-1} F(x_k, y_k) \Delta, \qquad (5.4)$$

and (5.3) by the recurrence relations

$$x_{k+1} - x_k = G(x_k, y_k)\Delta, \qquad x_0 = c. \qquad (5.5)$$

Here, we have set

$$x_k \equiv x(k\Delta), \qquad\qquad y_k \equiv y(k\Delta). \qquad (5.6)$$

In place of the original maximization problem, we consider the problem of maximizing $J(\{y_k\})$ over all N-dimensional sets $(y_0, y_1, \ldots, y_{N-1})$, where x_k are related as in (5.5). It is to be expected that as $\Delta \longrightarrow 0$, the maximum of $J(\{y_k\})$ will approach the maximum of $J(y)$, under appropriate assumptions concerning F and G. This is indeed the case.[7, 12, 13]

Now observe the similarity between the problem in this form and the multi-stage decision problem discussed in abstract terms in Section 4. The correspondence is as follows:

$$c \sim p,$$
$$y \sim q,$$
$$c + G(c, y)\Delta \sim T(p, q), \qquad (5.7)$$
$$F(c, y)\Delta \sim r(p, q),$$
$$f_N(c) = \text{Max } J(\{y_k\}) \sim f_N(p).$$

The basic recurrence relations are

$$f_1(c) = \underset{y}{\text{Max }} F(c, y)\Delta,$$
$$f_{k+1}(c) = \underset{y}{\text{Max }} \left[F(c, y)\Delta + f_k(c + G(c, y)\Delta) \right]. \qquad (5.8)$$

DYNAMIC PROGRAMMING   209

The numerical solution of the original variational problem has thus been reduced to the computation of a sequence of one-dimensional functions.

6. Examples – I.

Consider the problem of minimizing

$$J(f) = \int_0^T \left[x^2 + x'^2 + a_1 f^2 \right] dt \tag{6.1}$$

subject to

$$x'' + \lambda(x^2 - 1)x' + x = f(t), \qquad x(0) = c_1, \quad x'(0) = c_2. \tag{6.2}$$

The discrete version requires the minimization of

$$J_N(f_k) = \sum_{k=0}^{N-1} (x_k^2 + y_k^2 + a_1 f_k^2) \tag{6.3}$$

where

$$x_{k+1} = x_k + y_k \Delta, \qquad x_0 = c_1,$$

$$y_{k+1} = y_k + \Delta \left[f_k - \lambda(x_k^2 - 1) y_k - x_k \right] \qquad y_0 = c_2. \tag{6.4}$$

We have replaced x' by y.

Let

$$\underset{\{f_k\}}{\text{Min}} \ J_N(f_k) = F_N(c_1, c_2). \tag{6.5}$$

Then

$$F_1(c_1, c_2) = (c_1^2 + c_2^2) \Delta,$$

$$F_{N+1}(c_1, c_2) = \underset{f_0}{\text{Min}} \left[(c_1^2 + c_2^2) \Delta + a_1 f_0^2 \right.$$

$$\left. + F_N \left[c_1 + c_2 \Delta, c_2 + (f_0 - \lambda(c_1^2 - 1)c_2 - c_1) \right] \Delta \right]. \tag{6.6}$$

The numerical solution requires the tabulation of a sequence of functions of two variables, and a sequence of one-dimensional minimizations. It is still not a trivial problem in view of the limited capacity of most current machines. In return, however, we obtain not

210   NONLINEAR CIRCUIT ANALYSIS

only a solution of the original problem, but a solution of the entire class of related problems.

7.   Examples – II
       Let us now consider the problem of minimizing

$$J(f) = \int_0^T \left[x^2 + x'^2 \right] dt, \qquad (7.1)$$

subject to

$$x'' + \lambda(x^2 - 1)x' + x = f(t), \quad x(0) = c_1, \quad x'(0) = c_2, \qquad (7.2)$$

where $|f(t)| \leqslant k$.
       Proceeding as above, we wish to minimize

$$J_N(f_k) = \sum_{k=0}^{N-1} (x_k^2 + y_k^2), \qquad (7.3)$$

where

$$x_{k+1} = x_k + y_k \Delta, \qquad x_0 = c_1,$$

$$y_{k+1} = y_k + \Delta \left[f_k - \lambda(x_k^2 - 1) y_k - x_k \right], \qquad y_0 = c_2, \qquad (7.4)$$

over all sequence $\left\{ f_k \right\}$, satisfying the constraints $|f_k| \leqslant 1$.
       The recurrence relation corresponding to (6.6) is

$$F_1(c_1, c_2) = (c_1^2 + c_2^2) \Delta,$$

$$F_{N+1}(c_1, c_2) = \min_{|f_0| \leqslant 1} \left[(c_1^2 + c_2^2) \Delta \right. \qquad (7.5)$$

$$\left. + F_N \left[c_1 + c_2 \Delta, c_2 + (f_0 - \lambda(c_1^2 - 1)c_2 - c_1) \Delta \right] \right]$$

Observe that the constraint $|f| \leqslant 1$ which complicates the variational approach by classical methods actually simplifies the computational solution by these means, since it reduces the interval over which the minimum must be sought.

DYNAMIC PROGRAMMING   211

8.   Examples – III
     Let us now consider the problem of minimizing

$$J(f) = \underset{0 \leqslant t \leqslant T}{\text{Max}} \; |x(t)|, \qquad (8.1)$$

over all $f(t)$ satisfying $|f| \leqslant 1$.
     Proceeding very much in the same way as above, the numerical solution reduces to the computation of the sequence

$$F_1(c_1, c_2) = |c_1|, \qquad (8.2)$$

$$F_{N+1}(c_1, c_2) = \text{Max}\left[|c_1|, \; \underset{|f_0| \leqslant 1}{\text{Min}} \; F_N\left[c_1 + c_2\Delta, \; c_2 + (f_0 - \lambda(c_1^2 - 1)c_2 - c_1)\Delta\right]\right]$$

Details may be found in Reference 8.

9.   Bang-Bang Control
     Finally, let us discuss the problem of determining $f(t)$, subject to $|f(t)| \leqslant 1$, so as to minimize the time required to drive $x(t)$ into the equilibrium state. Consider the recurrence relations

$$x_{n+1} = x_n + y_n\Delta, \qquad x_0 = c_1,$$
$$\qquad (9.1)$$
$$y_{n+1} = y_n + \left[f_n - \lambda(x_n^2 - 1)y_n - x_n\right]\Delta, \qquad y_0 = c_2.$$

Denote this minimum time by $F(c_1, c_2)$. Then $F(c_1, c_2)$ satisfies the equation

$$F(c_1, c_2) = \Delta + \underset{|f_0| \leqslant 1}{\text{Min}} \; F\left[c_1 + c_2\Delta, \; c_2 + \left[f_0 - \lambda(c_1^2 - 1)c_2 - c_1\right]\right]. \quad (9.2)$$

From this equation we can calculate the isochrones, i.e., the curves

$$F(c_1, c_2) = k\Delta \qquad (9.3)$$

starting with the locus of the points on the curve $F(c_1, c_2) = \Delta$.
     The details of the most efficient computational procedure do not seem to be trivial. *

10.   Stochastic Control Processes
     In a variety of problems, there exixts an appreciable external influence of a type which can sometimes be considered to be "random". This is usually called "noise". In place of Eq. (2.6) we have an

     *Actually the dual process is easier to treat. We shall discuss this subsequently.

212   NONLINEAR CIRCUIT ANALYSIS

equation

$$\frac{dx}{dt} = g(x) + f(t) + r(t), \qquad x(0) = c, \qquad (10.1)$$

where $r(t)$ is a random function.

If $g(x)$ is linear, and we agree to minimize the expected value of the deviation

$$D(f) = \int_0^T (x - y, \ x - y) \, dt, \qquad (10.2)$$

we see that all that is required of $r(t)$ is a knowledge of

(a) $E\left[r(t)\right] = \bar{r}(t),$

(b) $E\left[r(t) \, r(s)\right] = k(t, s).$ (10.3)

The problem of determining $f(t)$ so as to minimize

$$J(f) = E\left[D(f)\right] \qquad (10.4)$$

is a variational problem that may be treated by the above techniques. A full account will occur elsewhere.

REFERENCES

1.   R. Bellman,   The Theory of Dynamic Programming , Princeton University Press (in press).
2.   R. Bellman,   Stability Theory of Differential Equations , New York, McGraw-Hill, 1952.
3.   R. Bellman,   A Survey of the Mathematical Theory of Time-Lag, Retarded Control and Hereditary Processes , The Rand Corporation Report R-256 (1954).
4.   R. Bellman, Dynamic Programming of Continuous Processes, The Rand Corporation Report R-271 (1954).
5.   R. Bellman, "Mathematical Aspects of Scheduling Theory", J. Soc. Ind. Appl. Math. (to appear).
6.   R. Bellman, "The Theory of Dynamic Programming", Bull. Amer. Math. Soc., Vol. 60, pp. 503-16, (1954).
7.   R. Bellman, "Functional Equations in the Theory of Dynamic Programming – VI, A Direct Convergence Proof", Ann. Math. (to appear).
8.   R. Bellman, "Notes on Control Processes – I, On the Minimum of Maximum Deviation", Quarterly Appl. Math. (to appear).
9.   R. Bellman, "On a Class of Variational Problems", Quarterly Appl. Math. (to appear).
10. R. Bellman, W. Fleming, D. V. Widder, "Variational Problems with Constraints", Annali di Matematica (to appear).

11. R. Bellman, I. Glicksberg, and O. Gross, "On the Bang–Bang Control Problem", Quarterly Appl. Math. (to appear).

12. W. Fleming, "Discrete Approximations to Some Continuous Dynamic Programming Processes", The Rand Corporation Research Memorandum RM-1501, June 2, 1955.

13. H. A. Osborn, "The Problem of Continuous Programs", The Rand Corporation Paper P-718, August 12, 1955.

14. N. J. Rose, "Theoretical Aspects of Limit Control, Report No. 459, Experimental Towing Tank", Stevens Institute of Technology, November, 1953.

DYNAMIC PROGRAMMING AND ADAPTIVE PROCESSES: MATHEMATICAL FOUNDATION*

by Richard Bellman and Robert Kalaba

In the previous work it has been automatically assumed, as in most of classical probability theory, that the random variables occurring had known distribution functions. In many interesting and significant processes, this is not the case. In addition to making decisions based upon the present state of knowledge, it is necessary, simultaneously, to learn more about the underlying physical system, either by means of actual testing and experimentation, or from observation of past performance, or by both.

In recent years, it has been realized that a large number of activities carried on in such diverse fields as control engineering, industrial processing, chemical engineering, economics, operations research, medical research, psychotherapy, design of automata and computers, and so on, involve processes of this complex nature. The mathematical treatment of these matters is complicated at the very beginning as a consequence of many vexing questions of formulation. Analytic and computational solutions are correspondingly more challenging.

In the paper that follows it is shown how the theory of dynamic programming allows the mathematical formulation of adaptive control processes along lines abstractly similar to the previous treatment of deterministic and stochastic control processes. Further results and extensive references will be found in:

R. Bellman, *Adaptive Control Processes: A Guided Tour*. Princeton: Princeton University Press, 1961.

---

* From *IRE Transactions on Automatic Control*, Vol. AC-5, 1960, pp. 5–10.

Reprinted from IRE TRANSACTIONS
ON *AUTOMATIC CONTROL*
Volume AC-5, Number 1, January, 1960

PRINTED IN THE U.S.A.

Dynamic Programming and Adaptive Processes: Mathematical Foundation*

R. BELLMAN† AND R. KALABA†

Summary—In many engineering, economic, biological, and statistical control processes, a decision-making device is called upon to perform under various conditions of uncertainty regarding underlying physical processes. These conditions range from complete knowledge to total ignorance. As the process unfolds, additional information may become available to the controlling element, which then has the possibility of "learning" to improve its performance based upon experience; *i.e.*, the controlling element may *adapt* itself to its environment.

On a grand scale, situations of this type occur in the development of physical theories throgh the mutual interplay of experimentation and theory; on a smaller scale they occur in connection with the design of learning servomechanisms and adaptive filters.

The central purpose of this paper is to lay a foundation for the mathematical treatment of broad classes of such *adaptive processes*. This is accomplished through use of the concepts of dynamic programming.

Subsequent papers will be devoted to specific applications in different fields and various theoretical extensions.

I. Introduction

THE PURPOSE of this paper is to lay a foundation for a mathematical theory of a significant class of decision processes which have not as yet been studied in any generality. These processes, which will be described in some detail below, we shall call *adaptive*.

They arise in practically all parts of statistical study, practically engulf the field of operations research, and play a paramount role in the current theory of stochastic control processes of electronic and mechanical origin. All three of these domains merge in the consideration of the problems of communication theory.

* Manuscript received by the PGAC, September 9, 1958; revised manuscript received, February 12, 1959.
† The RAND Corporation, Santa Monica, Calif.

Independently, theories governing the treatment of processes of this nature are essential for the understanding and development of automata and of machines that "learn."

We propose to illustrate how the theory of dynamic programming [1] can be used to formulate in precise terms a number of the complex and vexing questions that arise in these studies. Furthermore, the functional equation approach of dynamic programming enables us to treat some of these problems by analytic means and to resolve others, where direct analysis is stymied, by computational techniques.

In this paper, general questions are treated in an abstract fashion. In subsequent papers, we shall apply the formal structure erected here to specific applications.

II. Adaptive Processes

We wish to study multi-stage decision processes, and processes which can be construed to be of this nature, for which we do not possess complete information. This lack of information takes various forms of which the following are typical.

We may not be in possession of the entire set of admissible decisions; we may not know the effects of these decisions; we may not be aware of the duration of the processes and we may not even know the over-all purpose of the process. In any number of processes occurring in the real world, these are some of the difficulties we face.

The basic problem is that of making decisions on the basis of the information that we do possess. An essential part of the problem is that of using this accumulated

6   *IRE TRANSACTIONS ON AUTOMATIC CONTROL*   *January*

knowledge to gain further insight into the structure of the processes, using analytic, computational, and experimental techniques.

From this intuitive description of the types of problems that we wish to consider, it is clear that we are impinging upon some of the fundamental areas of scientific research. Obvious as the existence of these problems are, it is not at all clear how questions of this nature can be formulated in precise terms.

Particular processes of this type have been treated in a number of sources, such as the works on sequential analysis [14]; the theory of games [13]; the theory of multi-stage games;[1] and papers on "learning processes" [2], [3], [5]–[8], [11].

III. The Unfolding of a Physical Process

In order to appreciate the type of process we wish to consider, the problems we shall treat, the terminology we shall employ, and the methods we shall use, it is essential that we discuss, albeit in abstract terms, the behavior of the conventional deterministic physical system.

Let a system S be described at any time t by a state vector p. Let $t_1, t_2, \cdots$, be a sequence of times, $t_1 < t_2 < \cdots$, at which the system is subject to a change which manifests itself in the form of a transformation. At time t_1, p_1 is converted into $T_1(p_1)$, at time t_2, $p_2 = T_1(p_1)$ is converted into $T_2(p_2)$, and so on, with the result that the sequence of states of the system is given by the sequence $\{p_k\}$, where

$$p_{k+1} = T_k(p_k), \qquad k = 1, 2, \cdots. \qquad (1)$$

The state of the system at the end of time t_N is then given by

$$p_{N+1} = T_N(T_{N-1}(\cdots T_2(T_1(p_1))\cdots)), \qquad (2)$$

where p_1 is the initial state of S.

If $T_k(p)$ is independent of k, which is to say, if the same transformation is applied repeatedly, then the preceding result can be written symbolically in the form

$$p_{N+1} = T^N(p_1). \qquad (3)$$

The interpretation of the behavior of a physical system over time as the iteration of a transformation was introduced by Poincaré, and extensively studied by G. D. Birkhoff [4] and others. It furnishes the background for the application of modern abstract operator theory to the study of physical systems, as, for example, in quantum mechanics [12]. The idea of using this fundamental representation in connection with the formulation of the ergodic theorem is due to B. O. Koopman.

IV. Feedback Control

With all this in mind, we are now able to introduce the concept of *feedback control*.

See R. Bellman [1], ch. 10.

Supposing that the behavior of the system as described by the foregoing equations is not satisfactory, we propose to modify it by changing the character of the transformation acting upon p. This change will be made dependent upon the state of the system at the particular time the transformation is applied.

In order to indicate the fact that we now have a choice of transformations, we write $T(p, q)$ in place of p. The variable q indicates the choice that is made. Consequently, we shall call it the *control variable*, as opposed to p, the *state variable*. To simplify the notation and discussion, we shall assume that the set of admissible transformations does not vary with time.

If q_k denotes the choice of the control variable at time t_k, we have, in place of (1), the relation

$$p_{k+1} = T(p_k, q_k), \qquad k = 1, 2, \cdots, \qquad (4)$$

with p_{N+1} explicitly determined as in (2).

The associated variational problem is that of choosing $q_1, q_2, \cdots, q_N$ so as to make the behavior of the system conform as closely as possible to some preassigned pattern. We wish, however, to do more than leave the problem in this vague format.

V. Causality

Turning back, from the moment, to the deterministic, uncontrolled process discussed in Section III, let us note that the state of the system at time t_{k+1} is a function of the initial state of the system and the number of transformations that have been applied. Consequently, we may write

$$p_{k+1} = f_k(p_1), \qquad (5)$$

where p_1 is the initial state of the system.

For the sake of convenience, let us merely write p in place of p_1. Then, the function $f_k(p)$ is easily seen to satisfy the basic functional equation

$$f_{m+n}(p) = f_m(f_n(p)), \qquad m, n = 1, 2, \cdots. \qquad (6)$$

This is the fundamental semigroup property of dynamical systems.

VI. Optimality

With the foregoing as a guide, let us see if we can formulate the feedback control process in the same terms.

To illustrate the applicability of the functional equation technique, let us consider a finite process, of N stages, where it is desired to maximize a preassigned function, ϕ, of the final state of the system, p_N. This is often called a *terminal control* process.

The variational problem may now be posed in the following terms:

$$\underset{[q_1, q_2, \ldots, q_N]}{\text{Max}} \quad \phi(p_N). \qquad (7)$$

This maximum, which we shall assume exists, is again a function of the initial state, p, and the duration of the process. Let us then introduce the function defined for all states p and $N = 1, 2, \cdots$, by the relation

$$f_N(p) = \operatorname*{Max}_q \phi(p_N), \qquad (8)$$

where q represents the set $[q_1, q_2, \cdots, q_N]$.

Let us now introduce some additional terminology. A set of admissible choices of the q_i, $[q_1, q_2, \cdots, q_N]$, will be called a *policy*; a policy which maximizes $\phi(p_{N+1})$ will be called an *optimal policy*.

In order to obtain a functional equation corresponding to (6), we invoke the

PRINCIPLE OF OPTIMALITY: An optimal policy has the property that whatever the initial state and initial decision are, the remaining decisions must constitute an optimal policy with regard to the state resulting from the first decision.

The mathematical transliteration of this statement is the functional relation

$$f_N(p) = \operatorname*{Max}_{q_1} f_{N-1}(T(p, q_1)), \qquad (9)$$

$N = 2, 3, \cdots$, with

$$f_1(p) = \operatorname*{Max}_{q_1} \phi(T(p, q_1)). \qquad (10)$$

Further discussion, and various existence and uniqueness theorems for the functions $\{f_i(p)\}$ and the associated policies will be found in [1].

In this way, the calculus of variations is seen to be a part of an extension of the classical theory of iteration, and of semigroup theory.

VII. Stochastic Elements

In order to treat questions arising in the physical world in precise fashion, it is always necessary to make certain idealizations. Foremost among these is the assumption of known cause and effect, and perhaps, even that of cause and effect in itself.

To treat physical processes in a more realistic way, we must take into account unknown causes and unknown effects. We find ourselves in the ironical position of making precise what we mean by ignorance.

At the present time, there exist a number of approaches to this fundamental conundrum, all based upon the concept of a random variable. Building upon this foundation is the theory of games.

We shall discuss here only the direct application of the concept of stochastic processes, leaving the game aspects for a later date.

The theory of probability in a most ingenious fashion skirts the forbidden region of the unknown by ascribing to an unknown quantity a distribution of values according to certain law. Having taken this bold step, it is further agreed that we shall measure performance not in terms of a single outcome, but in terms of an average taken over this distribution of values. Needless to add,

this artifice has been amazingly successful in the analysis of physical processes; *e.g.*, statistical mechanics, quantum mechanics.

Following this line of thought, we begin to take account of unknown effects by supposing that the result of a decision q is not to transform p into a fixed state $T(p, q)$, but rather to transform p into a stochastic vector z whose distribution function is $dG(z, p, q)$, dependent upon both the initial vector p and the decision q. Let us further suppose that the purpose of the process is to maximize the expected value of a preassigned function, ϕ, of the final state of the system.

Before setting up the functional equation analogous to (9), let us review the course of the process. At the initial time, an initial decision q_1 is made, with the result that there is a new state p_1, which is observed. On the basis of this information, a new decision, q_2, is made, and so on.

It is important to emphasize the great difference between a feedback control process of this type, in which the q_i are chosen stage-by-stage, and process in which the q_i are chosen all at once, at some initial time.

In the deterministic case, the two processes are equivalent, and it is only a matter of convenience whether we use one or the other formulation.[2] In the stochastic case, the two processes are equivalent only in certain special situations. We shall be concerned here only with the stage-by-stage choice.

The analog of (10) is then

$$f_1(p) = \operatorname*{Max}_q \int_z \phi(z) dG(z, p, q), \qquad (11)$$

and that of (9) is

$$f_N(p) = \operatorname*{Max}_q \int_z f_{N-1}(z) dG(z, p, q), \quad N = 2, 3, \cdots.[3] \quad (12)$$

This type of process has been discussed in some detail [1].

VIII. Second Level Processes

Fortunately for the mathematician interested in these processes, the tale does not end here! For in a number of significant applications, it cannot be safely assumed that the unknown quantities possess known distribution functions.

In many cases, we must face the fact that we are dealing with more complex situations in which far less is known about the unknown quantities. For a discussion of the importance of these processes in the general theory of design and control, see McMillan [9]; for a discussion of the dangers and difficulties inherent in *any* mathematical treatment, see Zadeh [15].

---

[2] This corresponds to the choice we have of describing a curve as a locus of points or as an envelope of tangents.

[3] The descriptive version of this equation, when no control is exerted, is, of course, the Chapman-Kolmogoroff equation, the stochastic analog of (6).

A first attempt in salvaging much of the structure already erected is to assume that the unknown quantities possess fixed, but unknown, distribution functions. Regarding deterministic processes as those of zeroth level, and the stochastic processes described in Section VII as first-level processes, we shall refer to these new stochastic processes as *second-level* processes.

Although it is clear that we now possess a systematic method for constructing a hierarchy of mathematical models, we shall restrain ourselves in the remainder of this paper to the discussion of second-level processes.

IX. Additional Assumptions

Some further assumptions are required if we wish to proceed from this point to an analytic treatment. These are

1) We possess an *a priori* estimate for the distribution function governing the physical state of the system, which, until further knowledge is acquired, we regard as the actual distribution.

2) We possess a set of rules which tells us how to modify this *a priori* distribution so as to obtain an *a posteriori* distribution when additional information is obtained.

3) We possess an *a priori* estimate for the distribution functions governing the outcomes of decisions, which, until further knowledge is acquired, we regard as the actual distribution, and, as above, we know how to modify this in the light of subsequent information.

In this paper, we restrict ourselves to the case of known physical states.

In formal terms, our state vector is now compounded of a point in phase space p and an *information pattern*, $dG(z, p, q)$. As a result of a decision q_1, there result the transformations

$$p_0 \to p_1 \qquad \text{(observed)}$$

$$dG(z, p^*, q) \to dH(z, p^*, q; p_0, G, q_1, p_1) \text{ (hypothesized).} \quad (13)$$

On the basis of these assumptions, and considering a control process which continues in time as described in Section VII, we wish to pose the problem of determining optimal policies. For the first time, we are considering adaptive processes significantly different from those of the usual deterministic or stochastic control process.

X. Functional Equations for Second-Level Processes

As before, we introduce the function

$f_N(p; G(z, p^*, q)) =$ the expected value of $\phi(p_N, G_N)$ $\quad (14)$
obtained using an optimal policy for an N-stage process starting in state (p, G).

Depending upon the objectives of the process, only one or the other of p_N and G_N may enter into ϕ. Examples of both extremes abound.

Arguing as in the preceding sections, we see that the basic recurrence relation is

$$f_N(p; G(z, p^*, q))$$

$$= \underset{q_1}{\text{Max}} \int_w f_{N-1}(w; H(z, p^*, q; p, G, q_1, w)) dG(w, p, q_1), \quad (15)$$

for $N = 2, 3, \cdots$, with

$$f_1(p; G(z, p^*, q))$$

$$= \underset{q_1}{\text{Max}} \int \phi(w, H(z, p^*, q; p, G, q_1, w)) dG(w, p, q_1). \quad (16)$$

These equations are quite useful in the derivation of existence and uniqueness theorems concerning optimal policies, return functions, and in ascertaining certain structural properties of optimal policies [1], [2].

If, however, we treat processes which are too complex for a direct analytic approach, as is invariably the case for realistic models, we wish to be able to fall back upon a computational solution. The occurrence of functions of functions, *e.g.* the sequence $\{f_N(p; G)\}$, effectively prevents this.

XI. Further Structural Assumptions

In order to reduce the foregoing equations to more manageable form, let us assume that the structure of the actual distribution is known, but that the uncertainty arises with regard to the values of certain parameters.

At any stage of the process, in place of an *a priori* estimate, $G(z, p, q)$, for the distribution function, we suppose that we have an *a priori* estimate for the distribution function governing the unknown parameters. Again, a basic assumption is that this distribution function exists.

The functional equations that we derive are exactly as above, with the difference in meaning of the distribution functions that we have just described.

XII. Reduction from Functionals to Functions

We are now ready to take the decisive step of reducing $f_N(p, G)$ from a functional to a function.

It may happen, and we will give as an example, that change in the distribution function, from $G(z, p, q)$ to $H(z, p^*, q; p, G, q_1, w)$ is one that can be represented by a point transformation. This will be the case if G and H are both members of a family of distribution functions $K(z; \alpha)$ characterized by a vector parameter α. Thus, if

$$G(z, p, q) \equiv K(z, p, q; \alpha)$$
$$H(z, p^*, q; p, G, q_1, w) \equiv K(z, p, q; \beta), \quad (17)$$

the change from G to H may be represented by

$$\beta = \psi(p, \alpha, q_1, w). \quad (18)$$

Then we may write

$$f_N(p, G(z, p, q)) \equiv f_N(p; \alpha), \quad (19)$$

and (15) becomes

$$f_N(p; \alpha) = \underset{q_1}{\text{Max}} \int_w f_{N-1}(w; \beta) dK(w; \alpha). \qquad (20)$$

The dependence upon q_1 is by way of (18).

XIII. An Illustrative Process— Deterministic Version

Let us now show how these ideas may be applied to the study of control processes. Consider a discrete scalar recurrence relation of the form

$$u_{n+1} = au_n + v_n, \qquad u_0 = c. \qquad (21)$$

Here u_n is the state variable and v_n is the control variable. Suppose that the sequence $\{v_n\}$ is to be chosen to minimize the function

$$|u_N| + b \sum_{k=1}^{N} u_k^2, \qquad (22)$$

subject to the constraints

$$|v_i| \le r, \qquad i = 0, \cdots, N - 1. \qquad (23)$$

Although the precise analytic form of the criterion function is of little import as far as the present discussion is concerned, we have used specific functions to make the presentation as concrete as possible. Furthermore, the defining equation need not be linear.

This is a simple example of a deterministic control process. Introduce the sequence of functions defined by the relation

$$f_N(c) = \underset{\{v_i\}}{\text{Min}} \left[|u_N| + b \sum_{k=1}^{N} u_k^2 \right], \qquad (24)$$

where N takes on the values $1, 2, \cdots$, and c any real value.

Then

$$f_1(c) = \underset{|v_0| \le r}{\text{Min}} [|ac + v_0| + b(ac + v_0)^2], \qquad (25)$$

and for $N \ge 2$, the principle of optimality yields the relation

$$f_N(c) = \underset{|v_0| \le r}{\text{Min}} [b(ac + v_0)^2 + f_{N-1}(ac + v_0)]. \qquad (26)$$

XIV. Stochastic Version

In place of the recurrence relation of (21), let us introduce a stochastic transformation

$$u_{n+1} = au_n + r_n + v_n, \qquad u_0 = c. \qquad (27)$$

Here $\{r_n\}$ is a sequence of independent random variables assuming only the values 1 and 0. Let

$$r_n = 1 \quad \text{with probability } p$$
$$= 0 \quad \text{with probability } 1 - p. \qquad (28)$$

The quantity p is known, and for simplicity taken to be independent of n, although this is not necessary.

We now wish to minimize the expected value of the quantity appearing in (22). This is now a stochastic control process of the type described above in general terms. Call the minimum expected value $f_N(c)$. Then, following the procedures of Section VII, we have the relations

$$f_1(c) = \underset{|v_0| \le r}{\text{Min}} \left[\int_{r_0} [|ac + v_0 + r_0|] + b(ac + r_0 + v_0)^2] dG(r_0) \right]$$

$$= \underset{|v_0| \le r}{\text{Min}} \left[p[|ac + v_0 + 1| + b(ac + v_0 + 1)^2] + (1 - p)[|ac + v_0| + b(ac + v_0)^2] \right], \qquad (29)$$

and, for general N,

$$f_N(c) = \underset{|v_0| \le r}{\text{Min}} \left[p[b(ac + v_0 + 1)^2 + f_{N-1}(ac + v_0 + 1)] + (1 - p)[b(ac + v_0)^2 + f_{N-1}(ac + v_0)] \right]. \qquad (30)$$

XV. Adaptive Control Version

Let us now consider the adaptive control version. We are given the information that the random variables r_n possess distributions of the special type described above, but we do not know the precise value of p.

We shall assume, however, that we do possess an *a priori* distribution for the value of p, $dG(p)$, and that we possess a known rule for modifying this *a priori* distribution on the basis of the observations that are made as the process unfolds.

If we observe that over the past $m+n$ stages, the random variables have taken on m values of 1 and n values of 0, we take as our new *a priori* distribution the function

$$dG_{m,n}(p) = p^m(1 - p)^n dG(p) \Big/ \int_0^1 p^m(1 - p)^n dG(p), \quad (31)$$

a Bayes approach.[4]

Once we have fixed upon a choice of $G(p)$, the *a priori* distribution function at any stage of the process is uniquely determined from the foregoing by the numbers m and n. This simple observation enables us to reduce the information pattern from that of the specification of a number or vector, in general, plus a function $G_{m,n}(p)$, to that of the specification of three numbers, c and the two integers m and n.

In this way, we reduce the problem from one requiring the use of functionals to one utilizing only functions. This is an essential step not only for computational purposes, but for analytic purposes as well.

Let us then introduce the sequence of functions $\{f_N(c, m, n)\}$ defined once again as the minimum expected value of the quantity in (22), starting with the

---

[4] This is an assumption of the type called for in Section IX. Although reasonable, it is not the only one possible. There are analytical advantages in choosing G to be a beta distribution.

10   *IRE TRANSACTIONS ON AUTOMATIC CONTROL*   *January*

information pattern of m ones and n zeros, and state c. Then

$$f_1(c, m, n) = \min_{|v_0| \leq r} \Big[p_{m,n}[b(ac + v_0 + 1)^2 + |ac + v_0 + 1|]$$
$$+ (1 - p_{m,n})[b(ac + v_0)^2 + |ac + v_0|]\Big], \quad (32)$$

where $p_{m,n}$ is the expected probability using the probability distribution in (31), *i.e.*,

$$p_{m,n} = \frac{\int_0^1 p^{m+1}(1 - p)^n dG(p)}{\int_0^1 p^m (1 - p)^n dG(p)}. \quad (33)$$

For $N \geq 2$, we have the recurrence relation

$$f_N(c, m, n) = \min_{|v_0| \leq r} \Big[p_{m,n}[b(ac + v_0 + 1)^2$$
$$+ f_{N-1}(ac - v_0 + 1, m + 1, n)]$$
$$+ (1 - p_{m,n})[b(ac + v_0)^2$$
$$+ f_{N-1}(ac + v_0, m, n + 1)]\Big]. \quad (34)$$

In this fashion, we obtain a computational approach to processes with general criteria and an analytic approach to processes with criteria of particular type. A thoroughgoing discussion of the analytic aspects of the solution of processes of this nature described by linear equations and quadratic criteria will be found in a forthcoming doctoral thesis by Marshall Freimer. Previous applications of these techniques have been made in [2] and [3].

XVI. Sufficient Statistics

The fact that the past history of the process described in the preceding paragraphs can be compressed in the indicated fashion, so that functions rather than functionals occur, is a particular instance of the power of the theory of "sufficient statistics" [10].

Many further applications of this important concept will be found in the thesis by Freimer mentioned above.

In a number of cases, this compression of data occurs asymptotically as the process continues; *e.g.*, the central limit theorem. A number of quite interesting questions arise from this observation.

XVII. Discussion

In the foregoing pages, we have attempted to construct a mathematical foundation for the study of the many fascinating aspects of the field of adaptive control. In further papers, we shall discuss a number of complex problems which arise from this approach.

From the purely mathematical point of view, we are now able to contemplate a theory of continuous control processes of adaptive type, obtained as a limiting form of the theory of discrete control processes. A variety of significant convergence questions are encountered in this way.

Furthermore, we can construct a theory of multistage games on the same foundations.

Finally, the problem of computational solution is by no means routine, and there are a variety of interesting approaches based upon approximations in function space and approximations in policy space to be explored.

From the conceptual point of view, we must face the fact that there are many further uncertainties to be examined in the state of the system, in the observation of the random effect, in the transmission of the control signal, in the duration of the process, and even in the criterion function itself.

Bibliography

[1] R. Bellman, "Dynamic Programming," Princeton University Press, Princeton, N. J.; 1957.
[2] R. Bellman, "A problem in the sequential design of experiments," *Sankhya*, vol. 16, pp. 221–229; April, 1956.
[3] R. Bellman and R. Kalaba, "On communication processes involving learning and random duration," 1958 IRE National Convention Record, pt. 4, pp. 16–21.
[4] G. D. Birkhoff, "Dynamical Systems," American Mathematical Society Colloquium Publications, New York, N. Y., vol. 9; 1927.
[5] M. M. Flood, "The influence of environmental nonstationarity in a sequential decision-making experiment," in "Decision Processes," R. M. Thrall, C. H. Coombs and R. L. Davis, Ed., John Wiley and Sons, Inc., New York, N. Y.; 1954.
[6] M. M. Flood, "On game-learning theory and some decision-making experiments," in "Decision Processes," R. M. Thrall, C. H. Coombs and R. L. Davis, Ed., John Wiley and Sons, Inc., New York, N. Y.; 1954.
[7] M. M. Flood, "On stochastic learning theory," *Trans. New York Acad. Science*; February, 1954.
[8] S. Karlin, R. Bradt and S. Johnson, "On sequential designs for maximizing the sum of n observations," *Ann. Math. Stat.*, vol. 27, pp. 1061–1074; December, 1956.
[9] B. McMillan, "Where do we stand?" IRE Trans. on Information Theory, vol. IT-3, pp. 173–174; September, 1957.
[10] A. M. Mood, "Introduction to the Theory of Statistics," McGraw-Hill Book Co., Inc., New York, N. Y.; 1950.
[11] H. Robbins, "Some aspects of the sequential design of experiments," *Bull. Amer. Math. Soc.*, vol. 58, pp. 527–535; September, 1952.
[12] J. von Neumann, "Mathematische Grundlagen der Quantenmechanik," Dover Publications, Inc., New York, N. Y.; 1943.
[13] J. von Neumann and O. Morgenstern, "The Theory of Games and Economic Behavior," Princeton University Press, Princeton, N. J.; 1948.
[14] A. Wald, "Sequential Analysis," John Wiley and Sons, Inc., New York, N. Y.; 1947.
[15] L. Zadeh, "What is optimal?," IRE Trans. on Information Theory, vol. IT-4, p. 3; March, 1958.